ACID REIGN

And the Rise of the Eco-Outlaws

by Patrick Curran

Copyright© 2011 Patrick Curran
All rights reserved.

No part of this book may be reproduced, stored in a retrieval system, or transmitted by any given means without the written permission of the author.

ISBN: 9780983661900

Library of Congress Control Number:
2011908479

Disclosure

Author's Note:

This is a work of fiction but deeply rooted in the Cold War era on the Colorado Plateau. Nevertheless, many of the accounts are purely fictional. Some of the characters are fictional accounts of real people, some are totally fictional, and some are fictional composites of several people of that era. These characters are products of the author's imagination and should not be construed as real.

CONTENTS

Author's note...iv
Prologue...vi
Map..x
Chapter One: Red-Eyed and Rowdy...1
Chapter Two: Lizard Head Pass..11
Chapter Three: "Almost Heaven, West Virginia"..........................16
Chapter Four: Take Home Pay...25
Chapter Five: Lucy's Lounge..35
Chapter Six: Sunday Stew..44
Chapter Seven: Hot Water...52
Chapter Eight: Troubleshooting...59
Chapter Nine: Bake Sales and Bombshells....................................71
Chapter Ten: Catalytic Action..78
Chapter Eleven: Realigning the Stars..85
Chapter Twelve: Where the Hurleyburley's Done........................90
Chapter Thirteen: Texas Hold 'em...100
Chapter Fourteen: Bitah Honeezgai (Sick All Over)....................107
Chapter Fifteen: Fire in the Ore House.......................................115
Chapter Sixteen: Burnt Offering..125
Chapter Seventeen: Silver Creek...131
Chapter Eighteen: Town Hall Meeting..141
Chapter Nineteen: When the Land is Sick the People are Sick........153
Chapter Twenty: Ninety Miles of Bad Road...............................162
Chapter Twenty-one: Fire on the Mountain................................175
Chapter Twenty-two: Pray for Us Sinners..................................179
Chapter Twenty-three: The Dance...185
Chapter Twenty-four: The Big Boys are in Town........................195
Chapter Twenty-five: ½ Inch Socket on a ½ Inch Nut................201
Chapter Twenty-six: Breakfast in Hollywood..............................206
Chapter Twenty-seven: Waking the Sleeping Ute......................210
Chapter Twenty-eight: A Call to Action......................................219
Chapter Twenty-nine: Building A Coalition................................227
Chapter Thirty: Widows of Cove...243
Chapter Thirty-one: Making the Wind Blow..............................257
Chapter Thirty-two: Wake up Johnny...272
Chapter Thirty-three: The Outlaw Game....................................277
Chapter Thirty-four: Taking On the Man....................................291
Epilogue...306
Acknowledgements..311
Photo Credits..316
About the Author...319

Prologue

Nowhere is the wildness of the high country on better display than in the San Juan Mountains of Southwestern Colorado. For some, it is sacred ground to be shared and protected. For others, a remote mineral belt to be subdued and exploited. On occasion, both altruistic and acquisitive motives have been aligned, unlocking nature's bounty and man's goodness. But, there is the dark side.

Few white men had made their way into the San Juans until silver was discovered. In 1887, after years of prospecting, David Swickheimer finally hit pay dirt in Rico. Unlike the huge, well-funded corporations that followed, David was a wildcat entrepreneur with a limited bankroll. Convinced that high grade ore was at a lower level, he continued to drill. The final blast at two hundred and sixty-two feet uncovered a fifteen-inch blanket of silver ore.

In 1890, the U.S. Treasury generously responded to his discovery with the passage of the Stanley Silver Act, calling for the purchase of four and a half million ounces of silver each month to back up the currency. Over two thousand mining claims were recorded in the Rico area, as silver production soared from a hundred thousand ounces to six hundred thousand ounces in three years. The population grew to five thousand people, and Rico could claim twenty-three saloons, two churches, and a thriving, red light district. By 1893, two million dollars of silver were produced by a hardy lot of prospectors and miners, driven as much by adventure as the lure of wealth.

Just getting to Rico in 1890s was a major undertaking. Weary travelers arrived battered and bruised on ox-drawn freight wagons. The

PROLOGUE

Silver nugget discovered in the Enterprise Mine by
Dave Swickheimer and his crew in 1887.
(All photo credits listed on page 222)

journey took three days along a treacherous toll road that carved its way down deep canyons and over high peaks. They were greeted by a motley crowd of miners, teamsters, and cowboys meandering up and down Glasgow Avenue. Barking dogs, braying mules, and bar room pianos filled the air. Starry-eyed prospectors waited anxiously outside the claim's office, as dance hall girls blew them kisses from the saloon next door.

Perched at nine thousand feet, Rico was no place for the faint hearted. The mountain hydraulics alone could do you in; blizzards, avalanches, and broken pipes were normal in the winter, as were flooding, cave-ins, and rock-fall in the spring. If you died in the winter, you stayed in the ice house until the graveyard thawed.

With the repeal of the Silver Purchase Act on October 31, 1893, the first mineral boom was over, and the consolidation began. In 1898, the American Smelting and Refining Company acquired most of the smelting companies in Colorado, chanting their gospel of scale economy. Standard Oil Company, the first American trust, was a major investor. As the doors of mineral prosperity swung open, the doors of frontier independence swung shut. Two of the dreaded pillars of greed were now in play: near absolute corporate power and the opportunity for obscene profit. Meanwhile in 1898, Madame Curie in far off France isolated radium from uranium ore, a discovery that would have a major impact on the Colorado Plateau.

Rico was fortunate to have an abundance of copper, lead and zinc, as the world ramped up for WWI. Prosperity briefly returned, as it did during WWII. By 1941, Rico was a struggling company town, owned by The Rico Argentine Mining Company.

On August 1, 1946, nearly a year after the first atomic bomb was dropped on Japan; President Truman created the Atomic Energy Commission (AEC), authorizing the purchase of all the weapons-grade uranium they could buy. Thus began the Cold War and the greatest arms race known to man. The uranium boom was on. By 1953, hundreds of uranium mines were operating throughout the Colorado Plateau and the Navajo Nation. Sulfuric acid was needed for the extraction process, and as luck would have it, fifteen million tons of sulfur-rich, iron pyrite lay below Rico. In 1954, the Rico Argentine built a sulfuric acid plant a half mile north of town.

Some say we won the Cold War, but not without grave public health and environmental consequences. Rivers were polluted, the land was poisoned, and more than fifteen hundred white and Navajo uranium miners died of lung disease on the Colorado Plateau. None of them were buried in the Tomb of the Unknowns at Arlington National Cemetery. There were no flag-draped coffins, no drum rolls, no flyovers.

Please rest assured, this is not one of those tasteless broadsides against capitalism. Most of the good folks in Rico did conclude, however, that unbridled capitalism had a way of bringing out the worst in mankind. They also learned that this unsavory behavior was more apt to occur, when, what they called, The Four Pillars of Greed underpinned a business venture. Now, there is always a certain amount of lawlessness involved in any commercial endeavor. But when the four pillars of greed are aligned...

- Corporate Power
- Unprincipled Leadership
- Inept Oversight
- Obscene Profit

...well, well, then you have a sure-fire, money-making-machine—ching, ching...ching, ching. Order fades and chaos ensues.

But then again, an anomaly of Frontier Justice, call it a noble mutation, is that:
Educated Outlaws on occasion beget Benevolent Outlaws.

Consider, where would we be without the likes of Bonnie and Clyde, or Pretty Boy Floyd, or Geronimo or Crazy Horse? Alas, America would be overrun with educated outlaws. Frontier Justice was in full bloom on the Colorado Plateau during the Cold War. The Atomic Energy Commission did their outlawing under the banner of national security.

Acid Reign: And the Rise of the Eco-Outlaws is the tale of benevolent outlaws: Johnny and Roy, roughnecks from the coal camps of West Virginia, and Chee Benally and the Widows of Cove from the Navajo Nation. When Johnny and Roy hit town in 1958, the sky was brown, the river was yellow, and the lights never went off at the acid plant. They fought to clean up the Acid Plant, along with Chee and the Widows of Cove who fought against a mining industry that killed thousands of uranium miners. Who knows, a rising of eco-outlaws may just save the planet.

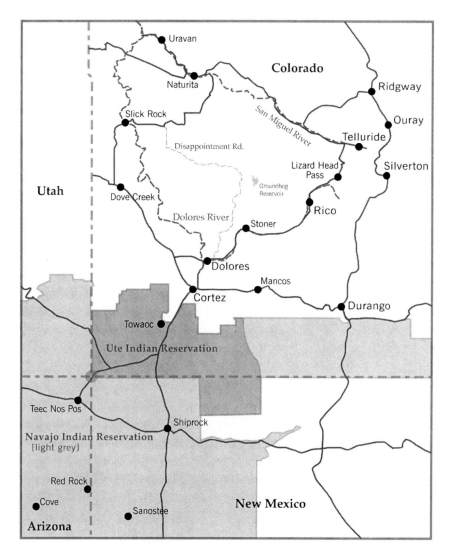

Map of Rico and Surrounding Areas

Chapter 1

RED-EYED AND ROWDY

Winter 1963

The Dolores River cuts a jagged path through the heart of the San Juan Mountains. Starting its two hundred and fifty mile run to the Colorado River in the silent snowfields of Grizzly Peak, then cascades through forests of spruce and aspen, and finally flattens out in the red rock country of the high desert.

The Rico Argentine Acid Plant is wedged on a narrow flat along the Dolores River, twelve miles south of its headwaters. The plant is a lone fortress of rusting warehouses, smoldering reactors and fuming stacks, bordered by pristine wilderness. It takes in iron pyrite from the Mountain Springs Mine and delivers sulfuric acid to the Atomic Energy Commission (AEC) mills on the Colorado Plateau, where it is used to produce weapons-grade uranium.

On a good day the Argentine was a money-making-marvel. This was not a good day. There was an emergency breakdown, a red alert in the high country. Gone was the plume above the stacks, and the deafening roar of the ore crusher. Three feet of snow surrounded the plant, and a light veil of spindrift swirled above the snow pack.

From the county road above the plant it all looked quite peaceful. Inside it was a belching inferno.

An emergency breakdown is like a collision at sea. An alarm bell blared and rattled through the metal warehouses. Johnny Carnifax, the acting foreman, raced to the control room and scanned the instrument panel. The reactor light flashed red. He grabbed the intercom, "All hands report to the reactor on the double, and bring your safety gear!"

As usual, Johnny and his buddy, Roy Loudermilk, got there first. The reactor was an enormous, cylindrical oven that rose forty feet above the snow pack. Three feet of bricks lined the inside. Crushed pyrite from the ore house poured into the top of the reactor, and when roasted at eighteen hundred degrees, produced a toxic vapor that eventually became sulfuric acid. If the reactor wasn't properly maintained, molten slag built up on the bottom and choked out the fire. This caused a breakdown.

Johnny and Roy clung to the top of a narrow railing that angled its way up the reactor to a manhole. Johnny slammed the handle of the spanner wrench against the iron railing to knock off the ice, and then stretched upward to slot it onto the last manhole cover bolt. Small irregular clouds of vapor pulsed from his facemask, and a stream of blood ran down his left arm and soaked the front of his shirt. Roy was wedged above him trying to steady the ten-pound sledge. He had missed the first time and shredded Johnny's hand.

"Now, damn it, now" Johnny yelled. The sledge fell hard on the spanner. The final bolt popped free, and the manhole cover shot across the yard. A cloud of acid fumes spewed out. The boys were tossed into the snow bank below the reactor. When Johnny opened his eyes, Roy was the only one around. The rest of the crew had vanished.

Roy rubbed his right hand. A deep cut ran across his forehead and down his cheek. He grimaced, as he mopped it with his shirt sleeve.

"Shit Johnny, I can't go on stage looking like this."

Roy was as vain as he was handsome. At six foot four with slightly stooping shoulders, he had a menacing, crane-like presence, which he used to full advantage. He and Johnny had grown up in a coal camp in West Virginia. The high point in Roy's twenty-two years on the planet was winning the state football championship his junior year of high school along with Johnny. They both quit school that year to work in the mines. Roy had lasted six months and had mostly drifted since then. He was no stranger to hard work, but did his best to avoid it. He was currently pursuing a singing career modeled after Elvis. A perfect night for Roy was taking on the meanest guy in the bar and then leaving with his lady. He called it a double header.

Johnny grinned fiendishly as he crawled out of the snow bank. His eyes were glazed and adrenaline roared through his body. He was of average height with broad shoulders and thick forearms. On the outside he was a brute, rock hard and lightening quick. On the inside he had his scars, suspicious of others, yet trying to do things right, but edgy, quick to snarl, like a dog that had been kicked too often.

"You got a ding on your head there, pretty boy."

"Come on Johnny, this sucks."

"Let's look inside."

Roy looked around, "The rest of your crew ran off."

Johnny shrugged and grabbed the sledge hammer. After six months of cleaning-up one mess after another, he'd been named acting foreman. Snow swirled outside the reactor and the gauge read thirty degrees. Inside the reactor, the gauge read two hundred degrees. He climbed through the manhole and disappeared into a searing tempest of ash and red fumes. A four-foot chunk of slag, called a clinker, blocked the air jets at the base of the roasting pyrite. It had to be cleared. He crouched on a narrow ledge and wedged his feet into the brick work, then leaned backward, extending his arms and began pounding on the clinker. This was a high-wire act not found in the safety manual. Far below puddles of molten ore bubbled on the reactor floor. Johnny braced himself. You didn't want to slip, he noted. Within minutes his facemask was clogged. He swore loudly and flung it out the manhole.

Outside, Roy stumbled around in the snow bank shielding his face from the biting wind. He finally found shelter under the railing that ran up the side of the reactor. Heat radiated from the iron structure. Muffled sounds of pounding and cussing came from inside the reactor. The blast collapsed Roy's bouffant and Dixie Peach Pomade ran down his neck. He went to work on it, getting the jelly roll right in the front and the ducktail to come together in the back. After several minutes he ambled back to the manhole and stuck his head in, "How ya doing there, fireball?"

Johnny gasped for breath. Even when you lived at nine thousand feet, you never got used to the thin air, and this was hot, toxic, thin air. He pursed his lips and inhaled slowly. His eyes were blood red, and his brows were singed. Yet he grinned, "We got us a big clinker. I got about half of it knocked out." Johnny's face was covered in ash, but his teeth gleamed white through his blackened lips.

"Man, you look evil."

Johnny held his grin, "Forget about the facemask, pretty boy. Take shallow breaths, it won't burn as bad."

Roy started shaking and bobbing. Johnny had seen this many times, sometimes outside a bar, and sometimes a thousand feet down in a mine. When Roy sensed danger, he went into a crisis mode, caught between fight and flight. In school they'd called him hyperactive.

"How about we let her cool off, get after her in the morning?"

"Grab the long bar, Elvis," Johnny said, shaking his head. "It's your turn."

Roy had learned not to cross Johnny when he was going full throttle. They'd been brought together by hard times in the coalfields. Roy won the bar fights and Johnny tried to keep him employed. They communicated in verbal jabs and punches, good-natured, but harsh sparring. Roy tossed his facemask in the snow bank and crawled inside.

"That's some cap ya got there."

The plastic bill of Johnny's hunting cap had melted away.

Johnny smiled dismissively, "Gone, ain't it? So look, hook one leg behind the pipe there and then brace your shoulder against the bricks, so you can free up your arms."

"Damn Johnny," Roy whined, "is this some sort of circus act?"

When he saw the molten slag on the bottom of the reactor, Roy's eyes lit up like the headlights on a snowplow. "What if I slip?"

"Then we'll send your ass home in a mason jar," Johnny hooted, jumping over the lip of the manhole and into the snowbank.

When he landed, the relief was immediate, yet numbing. He rolled on his back, and scooped snow on his face and arms. It stung, and then it soothed. The flaming slag paled and the hammering of the sledge faded. He drifted, tossed in a sea of darkness…a cloaked figure emerged, someone was knocking…it grew louder. "Wake up Johnny, don't you get it? Stop playing the hero."

It was his mother. Then it was Polly his wife, "Wake up big shot, wake up. Get smart, you can't do it all. A foreman gets things done through others."

"Johnny, God damn it, get up," Roy shouted as he jumped out of the reactor into the snowbank. He grabbed Johnny's shoulders and shook him. "You okay, man?"

Roy's face was blurred and his words were garbled. Johnny was spinning at the bottom of a pool, drowning, always drowning. He kicked violently toward the light, thrown back into darkness by the rush, and then he was back. Rubbing his head, he whispered, "Whoa, lost it there for a minute." He got to his knees, then crumpled, "Must have been the fumes."

Roy collapsed beside him.

Just then Chester Ratliff, the Mine Superintendent, roared around the corner and skidded to a halt. He was all angles, a square jaw, and a flattop, no one to fool with. Chester was torn between being a top-notch mineral engineer and a loyal employee, which is to say he needed the money. To his left, sulfur dioxide spewed out of the reactor, and to his right, Roy and Johnny lay crumpled in the snowbank, charred minstrels in shredded overalls and brimless hunting caps.

Chester bit off a smile as he disappeared into the manhole.

Johnny got to his feet and tried to steady himself. Chester crawled out of the reactor and silently studied his small black book. He measured his words carefully.

"It's two p.m. We need to light off the reactor by five, and you can bet your ass that Mr. Stanley T. Pritchard will be here at the stroke of five." Johnny and Roy nodded. "And Johnny, you need to round up that crew of yours. Rotate 'um in and out."

Roy started to object, but Johnny cut him off, "We're on it, Sir. We'll have her online by five."

Chester nodded and then he was gone. The boys were similar in some ways, but so different. They both liked being out there on the edge, but Johnny was a fixer, and Roy was a breaker.

"You're getting to be a real candy ass," Roy taunted.

Johnny shrugged, scooped more snow on his face, and stared blankly at his crew as they bolted on the manhole cover. He tried to piece together what had happened. He knew he had lost it for a while. In the coalfields, he had learned what they called "the edge", to deal with pain, to overcome fear, to stand alone. He had taken his share of lickin's, but along the way had mastered "the edge." Only the real bad guys messed with him, and he kind of welcomed that. But as Polly reminded him, "the edge" would get him killed, someday. In spite of that, he enjoyed fixing things. Yet, it troubled him. Most of what needed fixing at the plant was obvious, and it generally made good business sense to fix it. So, why did they let things run down?

Roy stepped back from the reactor and growled, "I'm startin' to feel like we're the canary in the coal mine. I say we toss a stick of dynamite in the reactor and head for California."

"You've always been a hit and run guy. I'd kind of like to make this work."

Johnny paused for a moment and then concluded, "I think I'll have a word with Chester once things settle down. There are lots of simple things we can do to get this place humming.

"What the hell is it with you and cleaning up other people's mess? And while I am at it, I think you said 'things we can do.' Who the hell is *we*? Cause you ain't the Lone Ranger and I damn sure ain't Tonto."

Then Roy shrieked, "What the hell!"

Johnny had slipped around behind him during his rant and had him in a bear hug. Roy's limbs flapped around like a chicken trying to fly.

"No, you damn sure ain't Tonto, you're Elvis, and you better '*stay off of my blue suede shoes.*' Of course, we're going to do it, along with my lame-ass crew."

At 4:50 pm, the reactor lurched into action. The temperature inside had risen to eight hundred degrees. Below, Johnny and his crew were sprawled out on the snowbank caught somewhere between heat exhaustion and hypothermia. Clouds of brown sulfur dioxide billowed from the main stack and drifted down into town. Mr. Stanley T. Pritchard, President of the Rico Argentine Mining Company, drove up to the reactor.

Lizard Head Peak

Chapter 2

Lizard Head Pass

January 1963
Where Nature Does as it Pleases

Sunday morning, two days after the emergency breakdown at the plant, the Spitzers and the Carnifaxs drove up to Trout Lake for the day. It was a clear and cold in the high country, and they were thrilled to be together. After hiking in to the cabin, they'd have lunch and try their hand at ice fishing.

Paul Spitzer was Polly Carnifax's older brother. And Dorothy Spitzer and Polly had been best friends since grade school. They'd all grown up in a coal camp in West Virginia. As they rounded the curve, a towering spire filled the skyline. The friendly banter faded. At thirteen thousand feet, Lizard Head Peak guards the entrance to the pass, a brooding presence, silently observing the human condition. Mother Nature clearly worked overtime when she made Lizard Head Peak, dominant, iconic, imposing, yet fragile as bone china, slowly crumbling.

As they drove on, they crossed broad grasslands lined with spruce trees. This gave way to granite boulder fields that climbed several thousand feet to the jagged skyline. Cirque lakes hung in every canyon. From any angle the view was stunning, a dazzling array of shapes, dusted with snow and capped by an ice blue sky.

Timberline marks a dramatic shift in habitat from the green serenity of subalpine meadows to the grey severity of alpine spires. Life is harder above timberline. The air is thinner; the slopes are steeper, and survival is a full-time job. Botanists like to make crisp distinctions between ecological zones. They miss the greater meaning, the cosmic irony of it all. Timberline is an amber light, a whisper in the wind with an Old Testament admonition: *Beware, you are in High Country.*

Polly and Dorothy chatted away in the back seat. Polly, smart as she was impulsive, was a natural beauty, wore her hair in braids and avoided makeup of any kind. She looked back at the spire. "Now I see it, the summit splits like the head of a lizard."

Dorothy, a willowy blonde wore horn rimmed glasses and looked quite bookish. Beneath the surface she was wildly eccentric, fascinated by astrology and the occult. Her husband, Paul, had his hands full.

"You know, the ancients believed that the lizard symbolized detachment, stepping back to gain awareness."

Johnny grinned. "Lizards ain't that smart."

"Lizards *are not* that smart," Polly shot back. She knew it was futile, but she couldn't resist chipping away at Johnny's rough edges. She turned back to Dorothy, "Tell me more."

Dorothy's glasses slid down her nose, as she looked back at the peak, "Well, to gain awareness, the Greeks felt you had to detach yourself from the day to day, to step back, to sit quietly in the sun like a lizard."

"Sounds like a union boss to me," Johnny quipped.

"Johnnnnneee," Polly chided. "It's a symbol, an idea. Sometimes, you need to move forward, and sometimes you need to step back, before you can move forward."

Dorothy looked at Polly and they both broke out laughing. Dorothy loved being with Polly, a sister-in-law and a lifelong friend. They both had married strong-willed men, and one of their pastimes was sharing strategies on how to housebreak them.

Dorothy whispered to Polly, "I don't think he liked the part about stepping back. They're full steam ahead guys."

That sent them into peals of laughter. Dorothy looked back at Lizard Head Peak, smiled and dozed off, gathering sacred shells on the celestial shore.

As they came over a rise, Paul pulled over, "See the tracks there?"

"Sure do," Johnny said.

"That's a section of the old Rio Grande and Southern Line. That's how they got ore out of the San Juans from 1891 'til around 1951. Over the years, there isn't a section of that line that wasn't torn out by avalanches or floods."

"How about that." Johnny loved a good story.

Paul's eyes lit up, "Helen Hicks, here in town, the Superior Court Judge, told me as a girl she rode the train to high school in Telluride. She said when they had a big snowstorm; they mounted a rotary plow to the front of the locomotive. So, picture this Johnny, you got this locomotive bucking through a wall of whiteness, and the rotary plow throwing a rooster tail of powder two hundred feet in the air, and an angry plume of black smoke from the engine shooting up into the sky. Then an enormous explosion of snow and steam, as the locomotive bursts through it all, and the track is clear. Now that's biblical."

Rotary Snowplow

Johnny nodded, "Amen, that's a damn fine story, sounds true. Now, would you say it was a battle of good and evil, or maybe, a story of man overcoming the forces of nature?"

"If you consider the white rooster tail as good, and the black plume as evil, well then it's clearly, a battle of good and evil, a sign."

"Excellent point there, Paul. However, I have always found this whole business of good and evil a tricky one. What's good for some is evil for another. Therefore, I'd say it was a story of man gaining a fleeting edge over Mother Nature."

Polly nudged Dorothy. They both held in their laughter.

Then Polly leaned forward, "Do you think you hillbilly philosophers could move it on down the road?"

"See what I mean, Paul? Here we are trying to raise the level of discourse, and they can't handle it.

In the front seat, Paul and Johnny discussed plans for the day. The temperature was in the mid-thirties, a warm day for January at ten thousand feet.

"I think we've got everything," Paul remarked. "Four sets of snowshoes if we need them. You and I can carry the backpacks with the rest of the gear."

Johnny was itching to get on the trail. "I've got the creel, the ice auger, and the poles."

"So here's the deal. We park at the trailhead and snowshoe to the far end of the lake. It'll take us about thirty minutes to the old trapper's cabin. From there we can ice fish or hike up to the falls."

"How do we stay warm out there on the ice?" Johnny asked.

"We don't, but this will help," Paul slid a half-pint in to view from his jacket.

"Paul, pull over!" Dorothy shouted.

A herd of elk grazed near the road. They trotted proudly up the ridge and out of sight, as the car came to a stop. Lizard Head Pass was a pleasant relief from the abuses of the mining industry. Unlike the mountains below, the pass was unspoiled by head frames and tailing dumps.

As they rounded the bend, Trout Lake appeared, an enormous field of ice skirted by green towering conifers. Minutes later they set off from the far end of the lake. Johnny insisted on going ahead to break trail and get the fire going in the cabin. He quickly left them in his tracks.

Chapter 3

"Almost Heaven West Virginia..."

As Johnny got higher up the trail, he tried to light-foot it across the crusted snow, but broke through and sank to his waist. He floundered in the soft powder, experimenting until he worked it out—heel slam through the crust compacting the powder, skate forward, and then high step. Once he put it together, he churned up the mountain, slam-skate-highstep, slam-skate-high step, a delicate balance of power and agility. If it got ten degrees colder, the crust would hold. Then it was slide-step-slide. If it got ten degrees warmer, he'd put on the snowshoes. He'd learned to move in the snow, as a boy, deer hunting with his dad. This native cleverness had kept him alive in Korea.

The wind rocked him back as he came over the ridge. He leaned into it, slam-skate-high- step, slam-skate-high step. Once into the rhythm of the movement, his mind wandered. Roy and Johnny had grown up in Quinwood, West Virginia. They lived in company housing, ate their first candy bars at the company store, and watched their first western at the company movies. There wasn't much to be happy about in the coalfields. Yet, they were happy, recklessly happy, thriving on the turmoil. Perhaps they didn't get it, perhaps they did.

West Virginia Coal Camp

    The coal wars of the twenties and thirties had embittered mine operators and miners alike. Battle lines were drawn and few moved off their version of the truth. Johnny and Roy had heard countless tellings of every strike and shootout. If your daddy was a coal operator during the Matewan Massacre, it was nothing more than a cold-blooded, unprovoked murder of seven, unarmed Baldwin-Felts detectives by striking miners. But if your daddy worked in the mine, you heard a different story. You were told that the hardworking miners of Bloody Mingo County went out on strike after years of abuse, seeking to form a union. "Two-gun" Sid Hatfield, Matewan's Sheriff and a perfect gentleman, was on the side of the striking miners. On May 19, 1920, a posse of gun-toting "Baldwin Thugs" road into Matewan and evicted the strike leaders from their company-owned homes. That's what it was, that's what triggered the whole damn thing.

    And then again if your daddy worked for a strikebreaking gang like the Baldwin-Felts Agency, you heard another version. Sheriff Sid was a twisted psychopath, and it was he who was on the union payroll, and it was he who shot Albert Felts in the back of the head as he was leaving town. And that's what started it all. The boys had also heard more than they cared to about yellow dog contracts, which forbid miners to join a union, and the Wagner Act of 1935, which allowed them to join a union. To talk about any of this publicly set off bitter arguments, if not a call to arms. In spite of the searing tension in the coalfields, mining coal paid far better than any other jobs. After an argument with their basketball coach, in their junior year, Johnny and Roy quit high school. The next week they were mucking coal. Johnny was drafted six months later and spent three years in the Air Force in Korea. He went back to coal mining when he returned. The pay had improved, when you weren't on strike.

Johnny stopped when he got to a clearing. The silence was cleansing. To the west he saw San Bernardo Mountain, and tucked behind it, Mount Wilson, one of Colorado's grandest fourteen hundred foot peaks. Towering above him to the east was the high rugged summit of Vermillion Peak. Its iron heavy cliffs burnt orange in the morning sun. He knocked the ice off his boots, his muscles aching, his chest pounding. Damn, he loved this stuff. He looked back down the trail. He was leaving a packed path behind him. Then he was off, slam-skate-high step, slam-skate-high step, again his mind drifted…the image of John L. Lewis appeared. Johnny had met him once at a union meeting in Beckley. He was the President of the United Mine Workers, and a hero to the mine workers. He had an iron grip on the union, calling for a strike whenever he needed to renegotiate a contract. In 1956, the mine operators agreed to a contract that increased the daily wage to $22.25 a day, big money at the time. The pendulum swung in favor of the mine workers.

Eventually, the coal operators had to increase their prices and watch the coal market give way to natural gas. Many of the smaller operators shut down. The bigger, absentee-operators continued to consolidate, mechanizing their operations, and displacing more and more miners. Many of the young folks moved on while their parents hung on in abject poverty. By 1957, Johnny got married. He had a job at the mine, but was only working three days a week. He had moved along from the mucking crew to the blasting crew, making a few more cents an hour. It was that fall that he got a phone call from Paul Junior Spitzer. He had driven west with his family several years earlier and got a job in Rico. He told Johnny they were hiring at the acid plant.

Out of the blue, several days later, Johnny got a call from Roy. Now Roy was another story. He went through jobs like a goat goes through tin cans. He got the heave-ho from the mine for punching the foreman after a cave in, and went on to sorting mail at the post office, but couldn't leave the help alone. After that he drifted up to Chicago, the last Johnny had heard.

"Hey Johnny, Roy here. Yeah, I am fine, living here in the southside of Chicago, working in the auto industry."

"Hey, pretty boy, good to hear from you. Tell me about the auto industry."

"I am into ac-cessories—tires, radios, auto parts, stuff like that."

"How does that work?"

"It's pretty technical. You, being a dumb ass miner, may not get it. You see I've got a complex organization here. You've got your acquisition that's critical. Then you've got your sales and distribution."

"Let's start with acquisition."

"You still on a party line? No. Ok, good, let's see, acquisition. The trick in acquisition is to find a car that's on a deserted street with parts you can fence, then it's a matter of getting her into our shop quickly, and back out on the street. Precise teamwork, that's what I am talking about."

"You're stealing cars?"

"Nope, we're stealing parts. It's been going real good, got me a gang, really tough guys from West Virginia, call ourselves the Ridge Runners. But now that I got you on the line, let me ask you, have you heard of grand theft auto?"

"That's stealing cars ain't it?"

"That's the thing, were not stealing autos, just parts."

"You can call them parts, the law calls them autos. You in jail or running?

"Is there another choice? Just kidding there Johnny, to be exact I'd call it running, you got to move with the times and all, but I am kind of stuck. I'm damn sure not coming back to West Virginia."

Three days later they headed for Rico. Johnny took one small bag. Most of his baggage had been pushed back into a dark corner of his mind. They made a short detour to Graceland and then headed west. Roy wanted to visit The King in his mansion.

Johnny had the shakes when he got to the cabin. His pants were frozen stiff and hung like stove pipes. He pulled off his gloves and worked the feeling back in his hands, then got busy splitting wood, opening shutters, and sweeping out the cabin. His feet were numb, but that would pass. When he got the shakes, he picked up the pace. When it came to exhaustion, there was no easy out. He just kept at it until the work was done or he collapsed. He knew he could out-suffer most people, it gave him an edge. Polly said he was cross wired when it came to pleasure and pain.

When Paul and the girls arrived, the stove was roaring, and water simmered on the stove. The cabin was a simple one room, log hut. The stove sat in the middle of the room, and a window looked out on the frozen lake. They huddled around the stove pulling off hats and coats.

Polly unpacked the picnic basket, "This is nice. Let's have tea."

Paul patted Johnny on the back, "Thanks for breaking the trail, big dog."

Within minutes, they were eating lunch and chatting about one thing and another, today's sermon, the beauty of the mountains, the dangers of ice fishing, settling in on the latest goings on from home in West Virginia. Paul poured tea for Polly.

She looked up sadly, "Today's Daddy's birthday."

"I know. I've been thinking a lot about him."

"Mom says they're doing fine. She's such a liar."

The stove was roaring. The tin roof groaned as it heated up. Outside, snow sloughed off the roof, and ice cycles dropped like daggers in the snow pack. Polly got up and moved to the window, hoping to hide the rush of emotions that overcame her.

Paul mopped the sweat from his face, "We send them money. They send it back."

Paul Senior was a big-hearted bear of a man, yet another victim of the coalfields. The day it happened, he was running the mine motor, pulling coal cars out to the tipple. A week earlier, his helper had been laid off. He stumbled when he ran ahead to switch the tracks, and lost both legs above the knee.

He was fitted with artificial legs, and the union gave him a job at the scale house, weighing coal trucks as they left the tipple. He had spearheaded the union election in the Quinwood operation and worked with the National Labor Relations Board on the final certification. He was a leader in the union movement. All had gone well until a fellow by the name of Billy Dumbro, a miner with a month's seniority, had bumped Paul for the scale house job. Billy's only handicap was laziness. Sure enough Billy got the job and Paul got the boot.

Neither the Spitzers nor the Carnifaxs had much love for coal companies, but they'd come to hate the union. Johnny took it the hardest. After that, he wasn't much for groups, he preferred to work alone.

The ever resilient Paul Sr. started running a poker game in the kitchen of a deserted house. His wife sold sandwiches and ran to the liquor store for the occasional pint. They made ten cents a rack on the pool table in the bedroom. Eventually, Paul tossed out the artificial legs, and got a wheelchair. His upper body was naturally powerful, but got even stronger pushing the wheelchair. On more than one occasion he vaulted out of the wheelchair onto the poker table and cold-cocked a cheater. He ran an honest game. They survived.

Johnny joined Polly at the window. "I know you miss him, I sure do. Why don't we send some money to my folks and have them buy groceries and stuff like that?"

She nodded faintly and wiped her eyes. Paul Jr. had started working for the Argentine two years ago, and had recently been elected a Justice of the Peace. He was the first to break the silence.

"So, Johnny I hear you're quite the hero up at the plant these days."

Johnny moved to the window. "You didn't hear that from my boss, did ya?"

"Word has it, you and Roy pretty much singlehandedly got the reactor back online."

"Is that what sent the foul smoke down our way?" asked Dorothy.

"Yes, dear, and that's what we were trying to fix."

Polly joined Johnny at the window. "When they say, singlehandedly?"

"Ah come on Polly," Paul grinned. "You know Johnny, he's fearless. I hear his crew just skedaddled."

"That ain't true, Roy was right there with me."

Polly smiled. "I see what you mean, singlehandedly."

Johnny's jaw tightened as he moved to the door, "We didn't have time to form a God-damn committee and vote on it."

"But Johnny, you are the foreman," Polly replied.

After an awkward silence, Dorothy intervened, "Paul, why don't you and Johnny go catch us some fish, we girls have a lot to catch up on."

Johnny slung the creel over his shoulder with a smile. Paul grabbed the poles and the ice auger, and they headed out on the frozen lake. An hour later they returned, frostbitten, yet triumphant. They were pleased with their catch, and their faces had a ruby glow that only comes from cheap whiskey.

Dorothy chuckled, "That rotgut works every time, doesn't it? Come on Paul, we're taking a hike up to the waterfall."

Polly sat next to the stove with a ball of yarn in her lap, "Come over here big boy."

Johnny pulled off his coat and joined her, "I think I know where this is headed, so let me say this, it ain't about me being no hero, they've gotta change the way their running things."

"And who's gonna change them?"

"Me and Roy, hell we showed 'um how to fix the clinker."

"Johnny, that's not true. You fixed it for them."

He put another log in the stove. "Most of the guys that work there are scared. They just wait to be told."

"You're supposed to coach them."

Johnny stared blankly into the fire, hoping the discussion was over.

Polly carefully tied off the yarn. "I am not feeling right about Rico, it's another company town. Most folks don't want change."

"Speaking of change, hasn't today been great?"

Polly smiled at his evasion. "Johnny, I am dead serious. You're not going to change your crew, by doing the work for them. And that's the end of that sermon."

"Praise God. So, what are you knitting?"

Polly arched her brow. "Speaking of change, you're going to be a father."

Chapter 4

TAKE-HOME PAY

"What you owe after deductions."

Dawn spilled over Dolores Peak like liquid gold, as Mary Margaret Fagan bounded up Glasgow Avenue through a blanket of new fallen snow. Mary Margaret, Maggie to her friends, was a big woman in full stride with flaming red hair streaming in her wake. A brightly colored Navajo shawl was wrapped around her shoulders and a silver squash pendent hung from her neck. A black velvet shirt fell to the ground, partially covering her hand tooled boots. She was a champion calf roper in high school and rode every morning. On a bad day she left her spurs on. Today was payday. She was early and angry. Her spurs jingled as she loped along. Final adjustments were needed before the checks were ready for signature by Mr. Stanley T. Pritchard, the President of the Rico Argentine Mining Company.

She turned into the Burley Building, kicked the snow off her boots and unlocked the door. Her second floor office was little more than a broom closet, yet it had a high eastern facing window, which she adored. The dank of wood rot and coal soot hung heavy in the room. She pulled up the lower pane and gazed out. Silver Creek rumbled down the mountain and slipped quietly into the Dolores River.

Maggie worshipped at many alters—the little Catholic Church on the hill, any high alpine meadow, sunrise on Dolores Peak. Her day began with a meditation at the window. Smoke billowed out of Jim Rychtarik's chimney up above Silver Street. Maggie smiled. Jim was the last of the pick and shovel miners. He rode Champ into the high country for months at a time, with Maybelline and Star, two retired army mules in tow, pannier boxes bulging, covered with canvas mantees, properly tied off. He was at peace in the wilderness prospecting for silver, working his claims, a quite man, deeply religious, but mean as hell if you crossed him. For those who enjoyed their solitude, he was a hero. Last Fourth of July, he waved down two dirt bikers racing up and down Silver Street, and politely asked them to hold it down. They nodded agreement, rode quietly down to the end of the street, and parked in front of a motor home with Texas plates. An hour later they were back at it, revving their bikes as they sped past his cabin. As the sun was setting, Jim grabbed his hatchet, walked calmly down the road, and began chopping the chrome spokes out of the dirt bikes. He was on the second bike when the Town Marshall finally pulled him off. Thereafter, he was known as Chopper Jim.

Above Jim's cabin, Maggie could see the summit of Dolores Peak at thirteen thousand feet. Ribbons of snow hung on the cliff bands that crisscrossed above timberline. Snow gullies cut across the bands and cleared a wide path through the timber all the way down to Silver Creek. You didn't want to be anywhere nearby when the heavy snow let go. Maggie lost herself in the moment, scanning the traversing lines of snow, searching for a deeper connection, a fleeting epiphany, something to get her through the day. She squinted and a blur of angular images flashed before her, slowly settling into a field of incandescent white crosses. Spiritually armed for the day, she crossed herself, closed the window, and got to work.

At eight o'clock sharp, Stanley T. Pritchard swaggered past her office, spun on his heel and returned to her doorway. "Good morning, Margaret," he said flatly. "I assume the paychecks are ready for my signature?"

At first meeting, it was easy to underestimate Mr. Pritchard. He was odd, and his get-ups were a great source of amusement. Maggie feasted over the details of each new outfit, but was careful he didn't notice. Today, he wore a wide brimmed fedora tilted slightly to the side, flared riding britches, and knee-high boots. The outfits changed from time to time, but not the eyes. They were dark eyes, set deep behind heavy lenses. They burned a hole through you like a laser when he was riled, and that was most of the time. Maggie, not easily quailed, flipped slowly through the paychecks.

"All but the Navajos sir, I am taking them over to the Rico Mercantile to make the adjustments, sir. The rest are on your desk, sir.

Stanley put his hands on his hips and leaned to one side for full dramatic effect. "Margaret please, do not get flippant. You know how I feel about this, business is business. They are good workers, yes, but they are Navajos, and they need to pay their bills." His eyes narrowed and his face flushed.

Maggie started to object, then thought better of it. She liked a good fight, but she picked her battles carefully. This was one she'd get to in due time. She had a certificate in accounting and her husband was the Lead Worker for the Colorado Highway Department in Rico. She needed the job, but not that much. "Is that it then, Mis-ter Prit-chard, sir?"

"Yes, that is it Mary Margret, that is ex-act-ly it." His voiced cracked as it climbed the scale. "I don't want a penny going out to the Navajos until they clear their accounts at the Mercantile."

Maggie turned away. *I don't want a penny going out to the Navajo, what an evil man.* Maggie had gone to school in the Four Corners area with lots of kids from the rez. They were regular folks like everyone else and good friends. She grabbed the phone before it could ring a second time. It was Lucy Fahrion.

"Good morning, Maggie," she cackled. "It's Lucy. The great one just strutted by. I am thinking Tom Mix with the hat."

Lucy's Lounge, like the Silver Dollar Bar in Jackson Hole, Wyoming, was the hub of revelry in Rico and right down the street from the Burley Building. Lucy also ran the projector at the Argentine movie house and was the local expert on western movies. She'd seen them all.

Maggie smiled, "Yes, his highness has arrived. The hat is definitely Tom Mix or Hopalong Cassidy, but the tight-ass, riding pants? Maybe Gary Cooper in High Noon? No, hang on, who's the famous director?"

"You're talking about Cecil B. DeMille? Yeah, I can see that, the boots definitely. So listen, I know it's payday, so behave yourself."

"Thanks, dear, can you come for dinner on Sunday after mass? Jim's back in town."

"I've heard. I'll be there with bells on, you be nice to Mr. DeMille."

Lucy knew everything that went on in Rico long before it became gossip. After her husband died in a rock fall in a Naturita mine, she bought The Metropole Bar and renamed it Lucy's Lounge. Miners came to drink, play pool, and moan. Cowboys came to drink, play pool, and fight with miners. The cowboys seemed happier. Wives came to gab with the girls and drag their husbands home. And the kids came on weekends for a hug and an ice cream. Lucy got along with everyone and knew where the bones were buried.

Maggie gathered up the Navajo checks and headed down the street to meet with Charlie England, who ran the Mercantile. Everyone ran a credit account, only the Navajos had their charges taken from their pay. As she entered, she nodded to Chee Benally, a big kid from the rez. His father, Norman, had worked at the Kerr-McGee uranium mines in Cove, Arizona for three years, then for the U.S. Vanadium in Uravan for twelve years. Last year he gave it up, when Chee landed a job at the Black Hawk Mine in Rico. They built a hooghan on the flat below the acid plant, and Chee did his best to take care of his ailing father.

Rico Mercantile, 1890s

Maggie dropped the checks on the counter, and gave Charlie a chilly smile. He had a hawkish nose and ears that stood out like door knobs, holding up an oversized green visor. A crisp white shirt and tie were tucked under his bib overalls. Six freshly sharpened pencils stood at attention in the pocket of the bib. He methodically cashed each check, deducting what he was owed. If anything remained, he slipped it into the pay envelopes. He looked up at Maggie and snickered.

"It's just good business."

"Get on with it."

Chee paced back and forth in front of the crackling potbelly, in a frayed Pendleton shirt and jeans with a big rodeo buckle. He too had been on the rodeo team in high school. The ring of the cash register cut through him. Charlie struck the no-sale key like a concert pianist; money moved out, and then back into the nickel-plated cash register. Maggie tapped her nails on the counter as her anger mounted. Outside, it had started to storm. Hail stones clattered off the windows. Charlie finally finished and slid the check-log across the counter. Maggie scribbled her initials on the log, turned to go, then suddenly turned back and grabbed the pencils from Charlie's pocket, broke them in two, and threw them into the air. "Stick that in your cash drawer, Mr. Mercantile."

"Now Maggie," he giggled, "Don't get emotional on me."

Maggie joined Chee at the stove, handing him his pay envelope. Her face was flush and her hair had taken flight. "He's a son of a bitch!"

Chee sighed, "Yes, he is."

She sat near the stove and rubbed her temples. At the counter, Charlie scurried around sweeping up the pencil fragments.

"How's Norman?"

"Not good, spitting blood again." Chee emptied the envelope on the floor, a charge slip and nine one dollar bills. He checked the numbers, "Most of it's for whiskey."

"You taking him to the doctor?"

Chee looked back at the bills, "I wanted to."

"Will the doctor at U.S. Vanadium see him?"

"Done that. Says it's from smoking."

"He never smoked."

"No, never. He's drinking more." Chee stared down at the charge slip, and was overcome. His mind drifted to his boyhood in Cove, racing ponies up Red Mesa, building a campfire at sunset, laughing at the moon. All were vivid memories of his father, carved like petro glyphs in red sandstone. Norman was strong, wise, loving, and now, coughing blood, fighting the pain, broken, chanting in the sweat lodge, fighting the evil spirits

"You can't blame him."

"I don't. I want to take him to the rez hospital down in Shiprock, but not with nine dollars."

Maggie rustled in her purse and pulled out a twenty dollar bill, "Get on down there. We'll work it out."

Chee shook his head. He started working in the mines when he was twelve, drilling blast holes with the big, gas-powered jackhammer. Most men couldn't lift it. The Navajos called him Big Thunder. He towered over Maggie, who was over six feet.

"I am not saying this again. Go on, take it."

Chee shrugged and took the money, "I guess this means I'll be chopping wood and shoveling snow for the next year."

"Make that two years," she smiled. "Be careful. Montelores Curve is iced up."

Chee smiled broadly and took the money. Maggie noticed that his hands were pale and shaking. She'd seen that before. The locals called it a hangover; doctors called it vibration syndrome. Chee hid his hands in his pockets and headed for the door.

Maggie was in a foul mood when she got back to the office. Head down, she stomped by Stanley's office striking her heels with full force, her spurs rang like a tambourine ensemble. Chester Ratliff, the Mine Superintendent, was with him, reviewing the weekly production report. Stanley sat at an angle behind an enormous oak desk, tilted back in his tufted leather chair, hands cocked behind his head, and his legs crossed in the nonchalant fashion he'd learned at prep school. From time to time, he leaned forward and delicately fingered the nail-headed trim on the arm chair.

Chester stood like a rock on the shore of Stanley's desk, in steel-toed boots and insulated canvas overalls, patched at the knees and smeared with mud and pyrite tailings. He'd been up all night clearing a cave-in at the Mountain Springs Mine, a few broken bones, but no fatalities. He turned to the tonnage report.

"It was a good week, we're back on standard."

Wielding his fountain pen like a stiletto, Stanley crossed out the one hundred and fifty ton standard in the report. "If you set the standard at a hundred and fifty tons a day, that's what you'll get. Hell, it was designed for that. Let's get it to one-seventy."

Chester felt the blood rush to his face. Almost imperceptibly, he started to rock from one foot to the other. He was a well-respected operator, who knew the mining business inside and out. Stanley, from a prominent Salt Lake City family, was a holy terror at controlling costs and making money. Thus far, the combination had worked. The acid plant was the cash cow of the holding company, and Stanley, the darling of the Board. As he reminded them at the last Board Meeting, "Rico is an outpost in the wilderness and I run it like the cavalry."

Chester held his ground, "We need to schedule downtime for maintenance, if we want to improve our output."

"Get it to one-seventy, and we'll talk maintenance. And another thing, look into adding a roaster to the plant. Hasn't been done before, but I think it could boost our output."

Bone-weary, Chester struggled to gain composure. Not the damn roaster again, he thought. The plant was well designed and didn't need to be jury-rigged. When it came to being stubborn they were equals, but that's where the equality ended.

"I don't know Chester, sometimes I don't think you get it. This is a volatile business. We need to strike while the iron is hot. Clev-er, com-mer-cial-ly clev-er, that's what we've got to be." He snickered, pleased with his turn of phrase.

Chester brushed a hand over his crew cut. "Right now, we're crowding the system. The reactor is backing up, the tailing ponds are overflowing, and the air here in town is rotten."

"Yes, yes, and right now, U.S. Vanadium will buy as much acid as we can make. That means we run the plant around the clock. Hell yes, crowd the system and the crew. There'll be time for maintenance, when the market cools off. You know, a tire rated at sixty thousand miles can go one hundred thousand miles, if you push it. Sure, you have to take a few risks, but that's what winners do. And don't worry about the fines from the Game and Fish. That's pocket change, and," he snickered again, "we've got friends in high places."

Chester pushed up against the front edge of the desk and stared down. "Here's what you need to understand. We use vanadium pentoxide pellets to convert sulfur dioxide to sulfur trioxide. Those pellets don't know from shit about being com-mer-cial-ly cleav-er. They just do their job. They convert, and when they are used up, they don't convert. Then, they clog the converter. Sulfur dioxide pours out the stacks and the conversion rate goes to hell along with the output of sulfuric acid. I'll be damned if that's commercially clever."

Stanley opened the next file on his desk and waved Chester off with a backhand, like he was swatting flies.

A Stately Pleasure Dome

Chapter 5

Lucy's Lounge

The snow came sideways in waves, as Johnny and Roy scrambled down the road in the gathering darkness. The tailing ponds were blood-red and brimming, surrounded by a blanket of snow. Acid fumes snaked off the ponds, drifted downwind, and disappeared in the falling snow. It was payday and the office closed at six.

By six-thirty, they sat at the back of Lucy's sipping their first beer, checks folded in their back pockets. Lucy's was one of Rico's cherished establishments. In a block of crumbling board and batten storefronts, the red sandstone structure stood out. Above a bay window on the street level, a gilded sign proclaimed that this was indeed, Lucy's Lounge. It had been the Metropole Bar in 1892. Lucy Fahrion bought it in 1950 or thereabouts, and upgraded it to a lounge, more of a family affair. On the second level, a chapel like peak rose over four arched windows, giving it a sense of propriety.

Most miners worked long hours underground and often lived in bunkhouses bolted to the side of a mountain. To them, Lucy's was a pleasure dome. The food was hot, the beer was cold, and the company was pleasant, before or after a fight.

Johnny slapped Roy on the back, "Cheer up, grumpy. You ready for another?"

For Roy, work was never good. It was just something you did, like eating your vegetables, something you bitched about. Roy had three main passions in life: fighting, sex, and singing. He had had his share of luck with the first two. Roy scanned the bar looking to satisfy one of his first two passions, "Yeah, you buying?"

Lucy's was dimly lit and cavernous. Trophy elk horns hung between portraits of well-fed, nude ladies. The long bar was lined with serious drinkers. An ornate mirror ran behind the bar, with an assortment of whiskey bottles on display. A brass cash register sat at the far end. Beside it was a Smith & Wesson revolver in a leather holster. Lucy called it, "the shit-stopper".

Bits of conversation floated down the bar—stray calves, frozen pipes, high school basketball, elk hunting, and fishing, always fishing. The nook across from the bar had been permanently commandeered by the women. Cushioned benches wrapped around a silver plated, wood stove. The ladies chatted happily, warming their feet on the stove rail. Lucy, a small woman with a big smile, was a blur of activity: shaking hands, taking orders, and hugging children. Periodically she erupted into convulsive laughter. People looked up, smiled, and then ducked back into their conversation.

Lucy made her way to Johnny and Roy, "Howdy boys, another round?"

"Yes, please," said Johnny.

"You boys hear about the broken water main?"

Johnny grimaced, "Again?"

"Yeah, Myron Jones and his crew are up there now. They gotta dig through three feet of frozen ground. Straight whiskey tonight."

Johnny downed what was left of his beer and smiled, "We'll do our best."

Roy wandered off swinging his head from side to side like a search light. A big busted lady had just come in the door, strutting her stuff. Roy reversed his course and headed in her direction.

Lucy sat next to Johnny and whispered, "You're a friend of Chee Benally, aren't you?"

"Big Thunder, yeah we hunt together."

"Then, you probably know that Norman, his father, is not well. Chee just called. He's down at the reservation hospital in Shiprock. Damn near had to hog tie Norman to get him down there. He doesn't trust white medicine."

"I heard. He took him to the medicine man last month."

"It's not good. So, today, the hospital is doing tests. Chee can't make it back tonight, needs someone to cover for him."

Johnny looked at his watch, "His shift starts in an hour. I got him covered. You got the phone number for the hospital?"

Lucy wrote the number on a page of her receipt pad and slipped it to Johnny, "Here, believe me Johnny, this is more than a cold. But that's another story."

Roy wandered back to the table.

"So?" Johnny inquired.

"She was looking for her husband, kind of. He's a mean looking son of a bitch, but I think I can take him. Either way it looks promising."

"Well, ain't you the savior of the frontier family."

Roy sat and they quietly sipped their beers. Johnny knew what was coming. Roy wanted to move on to California.

"Johnny, I am serious, it's wrong what's happening up at the plant."

Outside, fog was rolling in.

"Yeah it's wrong," Johnny said, "but it ain't illegal. Stealing cars, now that's wrong and illegal. We need to be smart about this. We ain't in no position to get fired right now."

"I've been talkin' to some of the boys at the plant, guys that have been around for a while. Get a couple of drinks in them and they'll tell ya some scary shit."

Roy pulled a scrap of paper from his pocket, "Listen to this stuff, stuff like leaking acid lines, explosions in the crusher room, and acid tailings running into the river."

He banged his beer bottle down with each grievance. By now the place was jumping. Several couples were spinning around the dance floor, and a fight had broken out at the pool table.

"Look, I ain't fighten ya on this," Johnny replied, "but we gotta be smart about it."

"Being smart and doing nothing is a lot like being scared, and that ain't like you, Johnny boy."

Just then Jim Starks walked up. He hauled iron pyrite down from the Mountain Springs Mine on the most treacherous road in the county. When fully loaded, the REO dump truck held twenty tons of ore and was about as stable as a one-legged pig. Jim looked spooked.

"Hey Jim, want a beer?"

"No thanks, had a close one last night."

Roy pulled a chair around for him, "You ain't looking your best there, Jim, what happened?"

"Ah hell it was my fault; I missed a gear coming around that first big hairpin below the mine."

"Is that the full story?"

"Well, that's the way they put it in the accident report."

"So what really happened?" Johnny probed.

"Well, don't let this get around. The report is right, I did miss a gear. Ya know, I'm on the day shift now, so Chester he calls me at eleven last night, nice and sweet like, says 'Jim, we need a load of ore to finish the shift. Can ya give us a hand?'"

Jim looked carefully around the bar and continued in a low tone, "Ya gotta be careful, we've got some snitches around here. So, the only truck that's running is the REO with all the bullet holes in it."

"Christ Jim, that thing's got no brakes," Roy grumbled.

"That ain't the worst of it, she's got no lights either. So, it's pitch dark, I've got a twenty ton load and no brakes. I miss the downshift on the hairpin, just didn't see it coming. I can't make the turn. I try to bank her on the uphill side. She gets up the bank and starts to roll back on the driver side."

By now, Jimmy is shaking badly.

"I believe I will have a beer. So, there I was with the REO up on the bank. Then she starts to tip. I jump out, and she rolls right over me, dammed if the cab door didn't stay open. Rolled over me like a cookie cutter, not a scratch on me. Can't sleep, can't stop shaking."

More beers arrived and they drank silently. After awhile Johnny asked, "So what did Chester have to say about all this?"

Jim grabbed the edge of the table to steady himself, "Well," he says, "You did the right thing there, Jimmy, when ya roll 'um, roll 'um up against the bank so you don't lose the load'. Not a God damn word about me nearly gettin' chopped in two."

—∞—

Outside a fog bank hung over Glasgow Avenue. You couldn't see across the street. A faint shaft of light cut through the fog at the Burley Building. It was seven o'clock and Stanley Pritchard was still hard at it. He stood at his mahogany desk, like a cleric at an altar. A map of the mine workings hung on the paneled wall to his right, an ant colony of seventy-two miles of tunnels with branching shafts and drifts on three hundred thousand acres. A productivity chart hung to his left. The plant was producing, on average, one hundred and fifty tons of concentrated sulfuric acid a day, if he fiddled the numbers a bit, or watered down the load. He toyed with his gold cuff links as he read a letter from the Vanadium Company of America.

"Dear Mr. Pritchard,

As you are fully aware, we now live in an atomic age. We are in engaged in a cold war that must be won. VCA has recently renewed its agreement with the Atomic Energy Commission to produce as much weapons-grade uranium as possible. This is more than a business opportunity, this is a patriotic duty. We are making every effort to expand our output of uranium ore here in Uravan.

Please submit a detailed plan with a timetable on how Rico Argentine can expand its current daily output of sulfuric acid to 225 tons a day. We value our relationship, but may have to consider other suppliers."

He gritted his teeth and stared at the productivity chart. The next moment an explosion rocked his office. A burst of light flashed across the wall. He instinctively ducked under his desk as plaster rained down, and his correspondence took flight. Shaking uncontrollable he searched for the phone. This can't be happening, he thought, not now, he gasped. They were relighting the reactor that evening. He pulled the phone from the rubble and called the plant.

"Hello Ted, what the hell was that?"

"Don't know sir, wasn't us. We cut the propane off twenty minutes ago. The reactor is back up."

"Then, what?"

"Don't know sir, think it came from town."

The explosion blew out Lucy's bay widow and rolled through the lounge like a rogue wave. Some stayed upright, clinging to a post or a pillar, others were flattened. Folks near the window got cut up a bit but nothing serious. An elk antler fell on the pool table ripping the felt. Mothers gathered their children and pushed toward the door. Men followed, most clutching their drinks.

Johnny dusted himself off, "Let's get out of here Roy."

Roy sipped his beer under the table, "I'm staying put, these things come in pairs."

Main Street was socked in. Most people lingered in a pool of light in front of Lucy's. A battered pickup roared out of the fog, screeched to a halt and Myron Jones jumped out. He winced as he saw the missing bay window. Myron was a lanky fellow with a drooping mustache and enormous hands. Rain or shine, he wore black rubber boots that came up to his knees, tattered denim pants, and a red flannel shirt. Tonight, they were all mud and ice. He owned a gold and silver mine in West Rico, and one up on Silver Creek, but only worked them when he needed pocket money. In his spare time, he served as the Mayor of Rico and the foreman of the maintenance crew. He was also a big hit with the Women's Club. They adored him. They thought he was the best looking man in Colorado and nominated him for County Commissioner every year.

"Everybody ok?" he grinned sheepishly.

Myron Jones, 1980

Lucy stepped out of the crowd, "A few scratches here and there. What the hell was that?"

Myron pulled on his mustache, "We'll have water by tomorrow, if I can find a ten foot section of pipe, down at the shop." He looked up the street with a smile, "The old pipe's on the roof of the church."

Lucy, who stood a foot shorter than Myron, started thumping him on his chest, "Myron I'm looking for straight answers, what happened?

Myron leaned against the fender of his truck and reflected, "Well, maybe one stick would have done it. We'd only gone a foot or so, after an hour of digging, ice, roots and boulders, a nasty combination. So I figured two sticks."

Lucy started cackling. A collective sigh of relief passed through the crowd.

"Hell, three sticks would have blown us off the mountain, Mr. Mayor."

Myron nodded politely, "It leveled Jim Rychtarik's shed. He came running out of his cabin in his long johns with a shotgun. Thought he was back in the war."

The water system was always a problem; the pipes had never been properly insulated. Rico drew its water from up on Silver Creek. Though it contained acid drainage from the mines, there was plenty of it. In the winter when the ground froze up, they kept the valve open so water ran continuously from the intake strainer on Silver Creek down to the storage tank in town. The overflow ran off into the Dolores River. If the strainer got clogged, they were in trouble. The water quickly froze and pipes burst. It seems, the new man on the maintenance crew had slept in, and hadn't got up to clear the strainer.

"That's pretty funny, but what about my window?" Lucy demanded.

Myron turned to the crowd, "The city will take care of all the damages. I'll see to that. I am sorry folks, but as I promise, you'll have water in the morning."

He jumped back in his pickup and disappeared into the fog.

Chapter 6

SUNDAY STEW

Ore buckets swung idly from the aerial tram that stretched from the Pro Patria mine to the mill in town. It was a lazy, cloudless Sunday. Dinner had been served yet the aroma of elk stew and biscuits still wandered out of the Fagan's proud clapboard house beside the tram. A large bay window with Victorian trim stood next to the front porch. Sunday was special in Rico. The clamor of commerce faded and the magic of the San Juans returned: the towering peaks, the roar of the river, the laughter of children. Brad Fagan, Myron Jones and Jim Rychtarik moved out to the porch, while the ladies sipped tea at the dining room table.

"You boys relax, and tell your tall tales. We'll have dessert and coffee in a bit," Maggie hollered from inside.

Jim sat on a stack of firewood, "Your Maggie makes a fine stew."

"Thanks," Brad replied. "That's the Irish in her."

Myron rocked back and forth on his haunches, "Where'd ya get 'um?"

Brad was the Lead Worker for the Colorado Highway Department. He and his brother in law, Michael Burns, took a week off each year to hunt.

"Way up on Telescope Mountain. He was a big bull, took two days to pack him out.

Brad pulled out a pint from the woodpile, "Well thanks, I'll drink to that."

Jim waved him off. Myron stopped whittling and had a drink, "Did I ever tell you boys about ole Buck, my mining partner?"

"I don't believe so," Jim said. He didn't hear so well and couldn't remember much anymore.

"Well Buck, you see, was what they call a mountain man. If he wanted an elk, why he'd sneak up and shoot it in its sleep. No chase there. One time he shot an elk, and he came back for it the next day. Well, a bear had drug it off and covered it up. That didn't stop ole Buck; he uncovered it and took it away on his mule."

"How about that?" Jim chuckled, "usually ole mister bear wins."

"You're right about that," Myron smiled, "So, tell me Jim, where's Maybelline these days?"

Maybelline and Star had become celebrities in Rico and a nightly topic at Lucy's Lounge, rated well above the weather or the latest fishing report. Maybelline started life as an Army mule with the Colorado Cavalry. She had a gentle disposition and the longest eye lashes in the army. When the cavalry was disbanded, Roy Pettingale bought her to pull the mine hoist at the Argentine. But Maybelline got pretty depressed, going round and round, day after day. So, big hearted Jim, bought her, and let her run on his pasture with his horses, Champ and Star. She was getting long in the tooth, but was still a good pack mule, if you didn't overload her. She got along well with the horses and had become best friends with Star.

Prospector heading for his claim.

Life was good on the pasture, until Helen Hicks, the Superior Court Judge for Dolores County bought Star and took her down to the Ascot mine. That broke Maybelline's heart. Jim had never had more than a single strand of barbed wire around his pasture and this was not enough to stop a broken-hearted mule. So Maybelline lies down near the fence and rolls under it, and off she trots down to the Ascot mine. This happened several times a week. The first question folks asked as they settled down for a drink at Lucy's was, "Where's Maybelline?"

"So, where is Maybelline?" Myron asked with a smile.

Jim pulled his chin, "I brought her back this morning. She's tied off."

"Hell, that's animal cruelty. You ought to be arrested."

As the bottle made its second round, they turned to the weather. Jim looked out on the north end of town, "Fine day, isn't it?"

Brad checked his watch. It was three-thirty, "The wind shifts around four these days, starts blowing down the mountain to the south. That's when it can get nasty."

Jim shook his head. He'd been out prospecting for the last two months, "Still at it are they?"

"Tell 'um Myron," Brad replied.

Myron continued whittling and looked down on the town, "Brad's right, on a bad day it turns dark around four, sulfur fumes blanket the town. Sometimes, you can't see across the street."

"What da the boys that work at the plant have to say?"

"Not much, if they want to keep their job," Brad replied.

Jim had drifted off, then his head jerked up and he looked around sheepishly, "A little cat nap there. Say Myron, was Buck the same fella that got gassed?"

"That's him. Ya see, his brother went down in the mine one morning, and he collapsed. He got some bad air or gas. So Buck goes down and drags him up and lays him out. Then Buck lies down and dies. The gas, plus the exertion of carrying out his brother, killed him. Strange thing was, his brother lived."

"Nice fella, that Buck."

"Yeah, I miss ole Buck. So Jim, how's the prospecting going?"

Jim paused for a moment to reflect, "Ya know, I am slowing down, but I've still got the fever. There's a fortune to be made up there. I got three new claims to stake next week."

Inside Maggie Fagan, Lucy Fahrion, and Trish Vanderville enjoyed their second cup of tea, chatting about family and friends. Lucy was born just up the road in Dunton, and lived on the Western Slope all her life. Trish was a new comer, a high spirited, social-activist from an old-line New York family. A mischievous smile beamed through her unruly, shoulder-length hair. She wore a red velvet jacket with a white cotton skirt that fell to the floor. To the dismay of her family, she had dropped out of Radcliffe in her junior year and moved out to Berkeley to join the free-speech movement. She graduated several years later, after spending time in jail for leading protest rallies. In September, she was hired as the fifth grade teacher in Rico. She also taught music and played a mean guitar. Her ambitions were modest; she simply wanted to change the world.

"So Lucy, where was your husband from?" Trish gushed.

"Jack was a miner, grew up in Hesperus west of Durango. We were married in Dolores, and he started at the Tomboy Mine in Telluride, then came down to Rico and worked in the Black Hawk Mine. When things got slow here, he went down to Naturita to work in the uranium mines. That was ten years ago."

She hadn't talked about Jack for a long time.

Trish was fascinated by the local history. "So, how did Jack get into mining uranium?"

Lucy got up and moved to the stove, "He followed the work. Things slowed down in Rico, and there was a uranium boom on the plateau. I think it was in 1946 that the Atomic Energy Commission started buying all the uranium they could get their hands on. Help me on this Maggie. You understand it better than I do."

"That's right honey, that's what triggered the boom."

Maggie jammed a log in the wood stove, "Well you know, we're in a Cold War and we need all the uranium we can produce to make those damn bombs. Every piss ant and soda jerk in America has a jeep and a Geiger counter nowadays, prospecting in the high desert."

Lucy went on, "Jack didn't go in for prospecting, he was a miner, worked for the U.S. Vanadium, or whatever they call themselves, had a great crew of Navajos working for him. Have you met the Benallys here in town?"

"I've seen the one they call Big Thunder," Trish replied. "He's a big fellow."

"That's Chee, and he's a fine fellow. His father, Norman, worked with Jack for years. They were great friends."

"So, how long did Jack work in the uranium mines?" Trish probed.

"Oh dear, how did we get here? Well, Jack was killed in 1953."

Trish set her cup down with a clatter. She felt a rush of blood to her face. "Oh Lucy, I am so sorry, I am so sorry. Oh dear, stupid me!"

Both Trish and Maggie moved to Lucy's side.

"Don't you worry girls, it's good for me to talk about it." Then she started to sob, "God, I loved that man."

They embraced silently.

Maggie headed for the kitchen, "Come on girls, let's make coffee and heat up the apple pie."

"Hold on, let me finish," Lucy threw up her arms.

She moved back to the stove and lowered her voice to a whisper, "A hanging slab fell on Jack. That's what killed him."

Maggie lowered her voice, "What Lucy's not telling you is." She laughed at herself and spoke up, "What the hell are we whispering about, the simple truth is those mines are damn dangerous."

"I'm not from these parts, but from what I hear, all mines are dangerous." Trish countered. She was a hungry predator looking for wrongs to right.

Over the years, Lucy had grieved quietly, never discussing Jack's death. She took off the gold locket from around her neck and opened it.

"That's Jack, he was a sweet man." Then she passed the locket around. No one said a word. When it got back to her, she looked at Jack and said, "Now Jack, today I am going to break that promise. They can't hurt me anymore, but they can damn sure hurt some of these other folks."

Maggie started clapping, "Let's hear it, sister!"

Lucy kissed the locket and fastened it back around her neck, "Well, hells bells, here I go. Now in Naturita, Jack said they had to get right back in the mines after they blasted, while the dust and smoke was still spewing out of them. He said you couldn't breathe, couldn't see a thing, nothing, not shaft damage, or hanging slabs. They had to get back in before any timbers were in place."

Trish pushed on, "Did Jack ever get to coughing after work?"

Lucy glanced at Maggie and winked. Trish was good-hearted but very direct. "Well yes, toward the end, he had a hacking cough and was often out of breath."

"Sounds like black lung disease to me, breathing all that dust and smoke," Trish insisted.

Maggie rolled her eyes. Her brother, Michael Burns, worked for the Public Health Service in Grand Junction, and had spent the last week elk hunting with Brad. Michael was well versed on mine safety.

"Hold on Trish," Maggie broke in, "If I am not mistaken, black lung disease comes from coal mines, this I think, is something else. Michael thinks it has to do with radiation. Brad says it's all very hush-hush."

Lucy frowned as she banged the stove door shut, "Well, whatever it is, it's awful, and Norman Benally has it. He can't stop coughing and he looks horrible. You wouldn't think the government would let this happen."

"It's the same thing that's happening here in Rico," Trish fumed.

"You're right about that sister," Lucy agreed.

Maggie smiled inwardly. She too had an agenda, and it was aimed at cleaning up the acid plant.

"Trish, I'd like you to join Lucy and me at the next Women's Club meeting. We plan to discuss mine safety and the pollution problems."

"I'll be there, if I can break away from school."

Brad stuck his head in the kitchen door, "Maggie, did I hear something about dessert and coffee?"

"Come on in boys, it's ready."

After dessert, Lucy stood up and folded her arms defiantly, "We've been talking about my Jack and Norman Benally. They worked together in the uranium mines in Naturita for years. You all know that Jack died in the mine, and most of you know that Norman is looking terrible these days."

Brad swallowed hard, "Do you really want to get into this Lucy?"

"Yes, Brad, I really want to get into this. I want to help Norman if we can. Something's not right about the uranium mines. And something is damn sure not right about the pollution here in town."

This was strong talk, even for this group. Lucy had planted a seed.

Chapter 7

HOT WATER

February 1963
"Always drink upstream of the herd."

Several inches of snow had fallen before dawn. For now it was clear and radiant. Polly took the leash off Storm, their Husky, once they crossed the highway. They scrambled across the train tracks and down the trail along the Dolores River. Storm raced ahead and Polly put on her snowshoes. When she saw the river, she called Storm back.

They hiked along the river every morning once she got Johnny off to work. Rambling down the mountain from the high country, the river left wetlands and beaver ponds on either side. This time of year the banks were covered in snow, and the river ran a lucid blue with clamorous white water sections at every drop. With hard rain, it turned milk chocolate, washing away the snow banks. Today it was mustard yellow, a hot luminescent yellow. Trout floated in the still water like fallen leaves.

Johnny had told her all along that the tailing ponds were spilling into the river. But this was more than the occasional overflow of the ponds or the slow leaching below them. It took thousands of gallons to turn the river yellow and etch the cobbled rocks along the banks. Sure as hell, someone had drained the ponds last night. Anger flooded though her.

The Argentine's official position was: "*a certain amount of leaching is to be expected. The river, you see, works like a water treatment plant, quickly diluting and purifying effluents.*" Tell that to the fish Polly fumed. There had been numerous complaints down river in the town of Dolores. Their tubs and toilets were permanently stained. The fish had disappeared and the cattle wouldn't touch the water. Every fisherman in the county knew you had to go above the acid plant to have any luck.

Polly and Storm kicked through the snow on the trail to the beaver pond, which was untouched by the acid tailings. Conifers ringed the pond, frosted with fresh snow. The morning sun glanced off the frozen pond, throwing splinters of light in every direction. Long ago, beavers had downed the smaller trees and dragged them off to build their dam and lodge. Storm sniffed every bush and quickly staked out her territory. Polly sat on a granite bolder and soaked in the natural beauty of the place. It had a narcotic effect, both energizing and soothing. This was her chapel in the wilderness, untrammeled by the hand of man. If nature had anything to teach us, surely it was in places like this.

Polly smiled; this was her place. She prayed that her pregnancy would go well, and that Johnny would find a way to work through the problems at the plant. She marveled at the way the beavers worked in harmony with nature. The wetlands they created with their dams enabled them to survive, yet helped control the flooding and created a new habitat. Surely humans could do as well. It was clear, she needed a plan.

Polly considered her options. She had pushed Johnny to become more of a foreman and to teach his men how to solve problems. Maybe that would do it, and Lord knows Johnny was good at fixing problems, when he wasn't cracking skulls. This was too big a problem to rely on only one solution. If the Argentine couldn't be changed, then they'd have to find a better place to raise their family. She tossed a stick over the snow bank, Storm cart-wheeled after it, lost in the deep powder.

The second part of her morning ritual was a visit to the Rico Town Hall, a formidable red sandstone and brick structure built in 1892. A one room library was located at the back of the first floor, where she read the local newspapers and browsed through any new books. Today was different. Today, she had a mission.

"Good morning Val," Polly said cheerfully as she entered the library.

"Morning," Val said flatly. She was a grim little lady with wire rimmed glasses and a tight bun of gray hair. She was as tough as a spent hen. Her husband, Alton, was on the skeleton crew at the Blaine Mine. The Argentine would ramp it up, if the market returned. Val worked part-time at the library. They got by.

Polly parked Storm by the stove for her morning nap and headed to the shelf where back issues of the *Dolores Star* were stacked. As she settled down at the reading table, she paused, and thought of her parents and the hardships they had endured in Quinwood, West Virginia. Every family had their stories: cave-ins, a crushed limb, black lung, shoot-outs with strikebreakers, death. Every child had heard the screaming, and seen the blood when they brought their fathers home. Polly covered her eyes. There was little that you could call justice in a coal camp, no permanent doctors, few honest sheriffs, and little concern for public safety. Yet, all coal camps weren't the same. Some had risen up. Some had gotten organized, not just unions, but a public outcry, a demand for better medical care and safer working conditions. It didn't happen often. But when it did, it was it was a sight to behold, people simply getting along and caring for each other. She'd seen it. She knew it could happen. That's what she wanted in Rico. It all started with a grass roots alliance. Amen. Today, she wanted to get the facts. To review the articles that had been written in the *Star* dealing with health and safety at the acid plant in Rico. After an hour, she had several pages of notes:

Dolores Star, Aug 1, 1958. Big explosion from leaking gas due to a faulty vaporizer. Had to replace the blower.

Dolores Star, Dec. 26, 1958. Due to trouble in the reactor the plant was down for 4 days.

Dolores Star, July 24, 1959. For the last two weeks the sulfuric acid leaking from the plant has been extremely bad, early Saturday morning it was so dense you couldn't see across the street.

Dolores Star, August 21, 1959. Since they removed the clinker from the furnace last week the acid plant seems to be working pretty well.

From across the room, Val eyed Polly suspiciously, as she jotted down her notes:

Dolores Star, October 2, 1959. L.B. Birch, superintendent of Argentine Mines, quit unexpectedly and returned to Salt Lake City.

Following a complaint from Dolores Mayor Bill Meyer to the Public Health Services on November, 1959. Rico Argentine agreed to double the number of tailing ponds and stop dumping the waste into the Dolores River. However, they refused to put liners in the ponds.

Polly shook her head in disgust. Your damn right the river is polluted. All you have to do is take a look for yourself. She made a note to take her camera tomorrow.

Val stood behind her for several minutes before Polly noticed. "Doing a little research we?" Val purred.

Polly looked up, surprised and flustered, "Well, yes, it's private research, do you mind?"

Val smirked and was gone.

Polly continued.

Letter from the Attorney for the Town of Rico to Mr. Stanley T. Pritchard, January 11, 1960:

"At the regular Town Board Meeting for the month of January, 1960, it was brought out that numerous complaints had been made to the Board by various citizens of the Town in regard to the conditions of the air in Rico caused by Rico Argentine's Acid Plant. For a few days last week the acid fumes and dust, etc. again created such dense fog over the town that an individual could hardly see twenty feet in front of him. Such matter in the air is harmful to humans and properties."

In response to these complaints, Miller and Parma, Rico Argentine's attorneys, had sent the town board a letter indicating:

"Every available precaution had been taken with that plant to guard against fumes and other disagreeable results and to provide for the safety of the workers and those in the immediate vicinity. Unfortunately, there was an occasion when, through human error, some inconvenience might have resulted to some of the inhabitants. You are assured that every available precaution will be taken to avoid any disagreeable or obnoxious results in the operation of the plant."

Yes, indeed Polly frowned, lawyers. Nevertheless, she was pleased with her effort. A case could certainly be made that the Argentine was negligent. She wasn't sure what to do with the information, but she planned to discuss it with Maggie Fagan. The pollution had clearly gotten worse in the last year. Polly gathered her things and went over to the front desk. "Bye now?"

Val's head popped up and she blinked vaguely, "Caught me cat-napping, anything else?"

"Do you have a phonebook for Southwest Colorado?"

"Over there, by the card catalogue."

Polly flipped through the phonebook, jotting down the numbers of the Dolores Mayor, the Dolores County Commissioners, the Dolores Chamber of Commerce, the Public Health Service, and the local Game Fish and Parks Department. That ought to do it, she closed her notebook and headed for the door, "I've got to run. Could you tell me when the Town Board meets next?"

"It's posted outside the Mayor's Office, *honey*."

"Well, don't let me keep you up."

"I think I know what you're up to young lady and you need to be very careful. Most of us complain in private, some complain anonymously to the Board, but damn few complain directly to the Board, especially those who work for the Argentine. I wouldn't rock the boat.

Polly tapped on her notebook, "Thanks, *honey*, and I hope this remains private, at least until I get out of the building."

"Aren't you the fresh one!"

Glenn Baer operating the three drum slusher in ore house

Chapter 8

Troubleshooting

The ore crushing was done in a huge, rusted-out warehouse. The wind slammed the loose metal sidings up against the studs on the outer walls, a harsh counterpoint to the roar of the crusher. Sulfur dioxide it seemed ate the nails first. The whole structure was supported by rough cut timbers. This was the dirtiest corner of the plant with the most primitive technology. Jaw crushers tore huge chunks of ore into gravel and cone crushers reduced them to granules. Sparks flew as the iron jaws struck the pyrite; the concussion shook the whole plant. A conveyer moved the crushed pyrite up a galley to the reactor. Dust hung like a storm cloud over the room.

Johnny pulled on his respirator. He was joined by Slim Webb who had been brought up from the mine as an electrician. They crossed the room and flipped the switch on the exhaust fan, nothing. The four foot blades were covered in spider webs.

"It's either the wiring or the motor," Johnny grumbled as he made a note in his pad.

"No, it could be the switch," Slim shouted.

"Like I said, it could be the wiring."

Slim was a bit of a wise guy. He'd bounced around for years. Word had it, he was an asshole buddy of the boss, a snitch.

Finding a problem before it became a problem was new for Johnny, harder. He was best at responding to an alarm bell or a flashing red light. Roy couldn't understand why he was working so hard at troubleshooting the plant. He figured that was the superintendent's job. But Johnny had made a commitment to Polly that he'd work at being a better foreman and he intended to do just that. He figured they weren't going to change things, unless he could show a cost savings.

Something bothered him as he observed the crushers. He pulled off his mask and scrapped the grit from his teeth. The filter was clogged again. An occasional spark flew from the jaw crusher. Something wasn't right. Crushed pyrite wouldn't burn unless it got really hot. No problem there, but he wondered, what about the wooden deck below the crusher? It was coated in lube oil and grime. He made a note.

"Could use a shield over the crusher," Johnny shouted above the roar of the crushers.

"Damn Johnny," Slim hollered, "it's freezing, and we got us a gale force wind blowing through here. You couldn't light a fire with a blow torch."

"Yeah, maybe so, but the wind's not always blowing. And it gets mighty dry in the summer."

He walked around the crusher, taking notes as he went. All the wood framing, the 2x4 planks, were caked with pyrite dust. Need to get this cleaned up. He moved on to the fire hydrant and hose. They looked to be in good working order. But if the plant lost power, they'd have to switch to an auxiliary pump down on the river to pump water up from the river. It was a bitch to get started in the winter. Better check on that.

They took a break. Slim waited inside, as Johnny went out to grab a smoke. He was frustrated by the problems, yet confident that most of them could be fixed. The evening air felt good. It had snowed again that afternoon, but had let up. He still wasn't sure he was cut out to be a foreman. He'd always been pretty good around machinery, but preferred to work alone. Something about the conveyor still bothered him. The reactor burned pyrite at 1800 degrees, but if it overheated, and if the conveyor was fully loaded, then burning pyrite could run back down the conveyor into the ore house, like a fuze on a stick of dynamite. Then he remembered something his dad had told him, "Be careful of any proposition with more than one *if* in it."

He pulled on his respirator and went back inside. He waved Slim over to the conveyor. With their masks on, they had to yell to be heard. Even then their voices were muffled. He told Slim his concern about a fire running back from the reactor into the ore house.

"You want my honest opinion," Slim bellowed, "I think this whole thing is a witch hunt. You're starting to sound like a union organizer to me."

In a flash Johnny grabbed Slim by the collar and jerked him up tight, "And how about you getting that fan working before I feed you to the jaw crusher?"

Slim spooked like a yearling. Polly was right; Johnny still couldn't control his temper.

As Johnny turned to go, he saw Chee leaning against the work bench, grinning broadly. He'd been there for awhile. They went outside to talk.

"You'd make a pretty good Indian."

"Damn right I would. So, how's Norman?"

"They did some tests, won't know anything 'til next week."

Chee patted Johnny on the back, "Thanks for standing in for me, brother."

Johnny looked up and smiled. Chee towered over him. For a big, mean looking guy, Chee could be very gentle.

"Anytime, I want to bring some elk meat up for you guys."

"Come see us. My father is in a lot of pain anymore, coughing fits and dizziness. We may have to move down the mountain. The altitude up here doesn't help. He sleeps off and on during the day, then not much at all at night. Says he's having dreams."

"What kind of dreams?"

"Bad dreams, Indians have bad dreams, ya know."

"Hell, I have bad dreams, too. Spooky, ain't it?"

"He says the bad wind attacks him. It comes in strange forms, screaming condors blowing him down the mountain, a pack of gray wolves circling around him, kicking up a whirlwind."

"What do you make of it?"

"Don't know. When a Navajos has these dreams, it usually means the spirits are angry. The dreams are a sign. But, the biggest sign is his sickness."

"Maybe the tests will help."

"Ha, there you go, talking like a white man, *bilagaana*."

"Anyway, I want to bring up some elk."

Chee wrapped his arm around Johnny's shoulder, "Ya know, he really likes you. He calls you Crazy Bear. He thinks you're crazy happy and crazy crazy. He likes that."

Johnny chuckled, "Crazy Bear hun? Hell, Polly would agree with that. You tell Norman he can call me whatever he damn well pleases. So, did they give him anything?"

"Not much, aspirin, cough syrup. They think it may be asthma. It's more than that though. And it's not just him. A lot of his buddies that worked together in the uranium mines are having problems as they get older, same stuff, trouble breathing, coughing, spitting up blood."

"We had black lung in West Virginia. Hell, I don't know. I'd better run. You tell Norman, I am happy with my Navajo name."

"Come in the afternoon if you can. He gets to drinking later on, and gets lost in his chanting and dreams."

He headed for the control room. Today, he was going to show Roy how to dump tailings. Polly was right, if he hoped to turn the plant around; he had to become more of a foreman. He met Roy outside the control room, "Hey buddy, let's get some coffee."

They pulled off their respirators and sat. Their faces were white where the masks had been. The rest was caked with grime. Roy coughed and spit red pyrite phlegm into his handkerchief.

"This mask ain't worth a shit."

Though required, respirators were a joke. They were designed to cover the mouth and nose, but never seemed to fit. The round filter on the side quickly clogged with smoke and dust. Even when the pads were replaced, they never worked for long, especially when the ventilation fans weren't working. The air was especially bad lately. The converter, which transforms sulfur dioxide into sulfur trioxide, needed an overhaul. Lethal levels of sulfur dioxide fumes poured out the stacks and blanketed the plant, production decreased, pollution increased.

Johnny smiled optimistically, "I got some ideas on how we can fix things around here."

"Whose gonna listen?

"I think they'll listen when it comes to cutting cost and making more acid."

Roy looked dejected.

"You ain't looking so hot there, Roy?

"I'm having problems with Ronda. You know, we ain't married or nothing but I thought it might lead to that."

"Hell, Ronda's an angel compared to Yvonne. What's up?"

"She sits in the trailer all day listening to soap operas. They make her sad, so she pounds down them sugar donuts."

"Sounds like Ronda to me."

"It makes her happy, and there ain't a hell of lot else for her to do. So, she's getting a real ass on her. I had to re-block the trailer, it tilts when she walks from side to side.

"Whoa, bad boy! Slow down there. Your mama would smack you good if she heard you talking that way. Let's just say, she cleans her plate real good."

"Well, okay, I went too far. Bad boy! So anyway, I've been reading about shock therapy."

"Really?"

"There was this great article in *True Confessions*, says some people need to be shocked into changing their way. I am thinking that's what Ronda needs."

"Ya gonna give it a try?"

"Yeah, well you know how religious she is. So, I told her that Saint Peter don't allow no fat ass people through the pearly gates."

"Hum? Well you're on solid ground there Roy. I think it was Luke in The Bible who said, '*make every effort to enter through the narrow door to reach salvation.*' So how'd it go?"

"Well, right off Ronda started screaming. So I figured the shock therapy was working. Then she said, 'her ass seemed to fit real nice in the backseat of my Plymouth the night we met.' And again I figured this was all part of the treatment."

"And?"

"Well then she kicked my bony ass out of the trailer. That's what she called it, bony. I think there's a song in it. Anyway, I slept in the back of Lucy's Lounge last night."

"So, any good news?"

"Well, the good news is I've got a gig at Dove Creek High School, for their homecoming dance. I'll be the singing lead, mostly Elvis stuff."

"That's good to hear."

Roy jumped up, swung his hips into motion, and started singing, *You ain't nothin' but a hound dog.*

"Easy boy, if Pritchard walked in here, you'd be riding your thumb to Nashville."

"Good, ain't it?"

Johnny scratched his head and grinned broadly, "Hang on, I'm having an e-piph-an-y, as the preacher man in Quinwood used to call it. Yeah, this is really good. Can you do, *Don't Be Cruel?*"

"Hell yes," Roy jumped up again and started to gyrate, *"Don't be cruel, to a heart that's true."*

"Sit down, what's the next line?"

"I don't want no other love, baby it's just you I am thinking of."

"That's it. So Roy, I am serious here now. You still love Ronda? I mean the whole package?"

"Hell yes. Ronda's ain't as touchy as Yvonne. Yvonne had a gun. Yeah, I love her, but..."

"But, nothing. Here's the deal. Get Ronda some kind of making-up present, not food. Then get down on your knees, on the doorstep of your trailer and sing that song. But don't do it, unless you mean it."

"On my knees? I don't know Johnny, what if she won't open the door?"

"Keep on singing 'til the dogs on your block start barking. Trust me, she'll let you in. Come on lover boy, we got things to do."

When the reactor was working, pyrite ore became sulfur dioxide and moved on to the scrubber. The slag fell down to the bed of the reactor. At least once a shift, someone had to pull the bed. This started by opening a valve at the bottom of the reactor. The molten slag then dropped down on a conveyer that fed to a slag bin. With Johnny's guidance, Roy had pulled the bed earlier in the shift. Now they had to empty the bin into the dump truck and haul it down to the tailing ponds.

As they got outside, a rising moon set off the jagged skyline. Snow drifted against the side of the reactor, sizzled, and was gone. The dump bin was overflowing with molten red slag. Johnny hated to waste time showing Roy, but this too was part of his plan.

"See that lever next to the bin?"

"I ain't blind."

"Ease it forward. Stay behind the shield."

The bin tipped down and slag poured into the bed of the Chevy dump truck. A spray of embers lit up the night. The slag quickly burnt off the snow that had collected in the bed of the truck.

"That's a load. Pull back on the lever and lock it off. That a boy. Now, get on down here."

They drove down to the tailing ponds.

"Why is this truck so rusted out?"

"Well now, metal conducts heat and paint burns at a certain temperature. Got the picture? This is the only truck on the mountain that doesn't need a heater."

"Johnny, you got to admit, this place is nastier than a coal mine."

The road to the tailing ponds ran along the Dolores River. The ponds were terraced on a flat that ran below the plant, one pond flowing into the next. The lowest pond spilled into the river.

"We just started dumping tailings on the ground rather than in the ponds," Johnny explained.

"How come?"

"They call it dry stacking, supposed to keep the ponds from running over."

"Hasn't done much good, has it?"

"Up here, they call it public relations, back home we'd call it bullshit."

Johnny pulled the truck up to an open area above the first tailing pond.

"If the wind is blowing, keep the windows up, or you'll get fried. It has a hydraulic lift. So once you're in position, just pull this lever and the bed rises and out she goes."

A steaming pyramid of slag formed behind the truck.

"Any questions?"

"More hazard, no hazard pay."

"Ya see that steep section of road coming down from the highway?" Johnny pointed across the river.

"Yeah."

"When it gets iced up bad, you need to dump a load of tailings on it. But you want to make damn sure you're going uphill when ya do it."

"Uphill, downhill what's the difference?"

Roy got nasty at the end of the day.

"Slim Pecker-Head Webb burnt the tires off the REO when he made a dump going downhill. The slag burned right through the ice and then it started running down the road. Damn sure fried them tires."

"Maybe you ought to be the one doing all this hazardous dumping."

"No, you'll be doing it from now on, and Roy don't be giving me a hard time about this. If we're gonna get anything changed around here, I'll need your help."

"Aye, aye sir, but I still think you're losing it." As they headed back to the plant Roy noticed a chute running along the road.

"So what's that?"

"That's a wooden flume, acid slurry from the scrubber runs down to the ponds. Got to be wood, the acid would eat up a metal flume."

"It's running pretty well."

"Yeah, that's why the ponds are running over. It's a shell game."

"And you think you can change these guys?"

"Damn right, "Johnny grinned, "why don't you get on out of here? You got your own problems."

"I'm gone. You're sure about the singing on the steps thing?"

"It works every time. You know Polly joined the Rico Women's Club. She really likes it. They're always busy, do good things around town. Ya think Ronda would be interested?"

"I'll check if she lets me. It might get her off them sugar donuts."

Women's Club on a Sunday Outing

Chapter 9

BAKE SALES AND BOMBSHELLS

Johnny left before dawn. He was determined to figure out how to fix the converter problem. Polly spent the morning preparing for her Women's Club presentation and was late for the meeting. She ducked inside the Town Hall and hurried up the stairs. Storm thought this was great fun and bounded after her.

Maggie Fagan, the club president, cut an imposing figure at the podium with her riot of red hair falling to her shoulders, complimented by a Navajo blanket dress of back and red design with a turquoise conch belt.

Storm raced past Polly and skidded to a halt at the base of the podium. Maggie who loved baseball, grinned, "I'd call her safe. What do you think ladies?"

No one challenged the call.

"Say hello to Polly Carnifax and Storm," Maggie announced. "Polly is our newest member."

Polly blushed as she searched for a chair. "Sorry. I thought it was at the school."

Stormed circled and curled up at her feet.

Maggie handed her an agenda and whispered, "You're fine. We just got started." She returned to her report. "So, as I was saying, we made five hundred dollars at the bake sale last week. Congratulations to all of you that worked on it. Half of that will go to the student scholarship fund and the other half will go to the summer swim program."

The cramped meeting room was tucked at the end of the hall. It also served as the bride's changing room for Town Hall weddings. A blue banner ran across the front wall announcing: Rico Women's Club in red appliqué. Another banner pinned to the sidewall proclaimed: "Let the Women Vote! They Can't Do any Worse Than the Men Have."

Polly nodded her approval, *Amen*, to that. In 1893, Colorado had been the first state to give women the right to vote by popular election. Women's Clubs throughout the state had rallied to the cause and were still the moral backbone of most communities. Women always played a major role in Rico's history. Dave Swickheimer, who put Rico on the map, was near bankruptcy in March 1887. Down over two hundred feet, he stopped drilling at the Enterprise shaft. Laura Swickheimer, his wife, saved the day. She purchased a $1 ticket in the Louisiana lottery from one of her boarding house tenants and won five thousand dollars. They paid their debts and kept on drilling. On October 6, 1887 at 262 feet, they uncovered a fifteen-inch blanket of silver ore. The silver boom was on.

Rico's own Betty Pellet, whose husband ran the Pro Patria Mine, was elected to the Colorado House of Representatives in 1940 and later went on to Washington. In the late 1950s, Helen Hicks became the Superior Court Judge for Dolores County and was instrumental in enacting legislation that outlawed wage garnishing of Navajo miners. Lately, most members of the Town Board were women. Some would say, especially in this gathering that they were doing a damn site better than the men had done.

Polly scanned the group. She recognized Lucy Fahrion and then Trish Vanderville, who was also a new member. Mary Joe Engel sat behind them, along with Sue Clark and Val Brown, the librarian. There were several others that she didn't know.

Maggie worked her way down the agenda: a Sunday outing with families to Dunton, new books for the library, and a workday at the cemetery in the spring. This involved cleaning up the tree fall, cutting back the brush, and setting up toppled gravestones. They also had to plan for the Fourth of July. Would they, or wouldn't they, enter a float this year? And, who would head the committee for the Old Timers' Picnic? All items were roundly discussed, brought to a vote and passed with ease. They agreed on most things. Silver-headed seniors in tight perms, mixed comfortably with younger women with shoulder-length hair. After an hour, they finished the new business and were ready for dessert. Any special presentations or speakers would follow. They ended the afternoon with several tables of Five Hundred.

"Thank you ladies, we're moving along nicely; let's break for dessert." Maggie announced and headed for Polly, "You're up next, Sugar."

She introduced Polly around to the other women during the break. "Now, you may not have met Elizabeth Pritchard and Mary Ratliff, wives of the muckety-mucks at the Argentine."

Oh my, thought Polly. Everyone was most gracious, glad to be out of the house and socializing. It had been a long, cold winter. The room was buzzing with pleasant conversation when Maggie called them back to order, "Hell…o…ho, we've got one more item on the agenda. I've asked Polly Carnifax to make a short presentation on a serious issue." She paused and lobbed a bombshell into the sisterhood, "Namely the pollution problem."

The room fell silent.

Val Brown gasped and whispered to Sue Clark, "Hitch up your knickers, Sweet Pea."

Polly knew that both their husbands worked for the Argentine and directed her talk to them. "Good morning. I know I am a newcomer, but if you will permit me, I'd like to share some information I've gathered." She opened her notebook, scanned the group, and considered suicide, but smiled grimly and plunged on.

"Every day, Storm and I walk along the Dolores River, and every day we see the same thing." She hesitated; most of the group had gone green, a lime Jell-O green. Maggie, her mentor and Stanley Pritchard's secretary, sat quietly to her left with a mischievous grin pasted across her face. To her right, Trish was shaking off her chair with excitement.

"And every day," Polly continued, "I see dead fish and yellow river rocks." She paused again and glanced at Maggie, who smiled with encouragement. "So, I've done some research. I'd like to start with articles from the *Dolores Star*. They document a series of incidents during the last three years, dealing with unsafe working conditions, air pollution, and toxic runoff from the tailing ponds."

By now, Elizabeth Pritchard and Mary Ratliff, who crowned the social registry in Rico, steamed like teapots. Their jaws clenched with their eyes wide open.

Elizabeth Pritchard raised her hand to speak, then strutted to the podium. She pursed her lips and was off in a most stilted cadence, "Pol-ly dar-ling, you are indeed a cour-age-ous wom-an, but I think ill-informed. Cer-tain-ly some of these things have occurred; though grossly ex-agg-er-at-ed by the Star. But, what you may not know is that a lot of time and effort has gone into co-rrec-ting these problems, so that aaaall." She paused and spread her arms wide. "Aaaall of us, can live the good life here in Rico."

A round of applause followed.

Maggie gagged.

Polly bit her lip, and then replied, "Mrs. Pritchard, ma'am, with due respect, these same issues have been voiced to the Town Board. The Argentine's lawyers have responded in writing, and I quote 'every available precaution has been taken to guard against these occurrences.' And yet, these occurrences have continued for the last three years."

Mary Ratliff, a silver-haired saint, rose to speak, "Polly, I am pleased that you have joined us. Our little club has done some wonderful things over the years, but I am not sure where you're headed with this."

"Thank you Mrs. Ratliff. Oh, Lord, I'm not either. I had hoped that this group, the women of Rico, would take a stand on this." Polly looked around the room.

"Yes, dear," Mary said kindly. "Do you have a motion?"

"I had hoped for some discussion."

"You've had it darling," snapped Val Brown, never one to miss an opportunity to ingratiate herself with the muckety-mucks. "Please state your motion."

"I think, excuse me, I move that the Women's Club petition the Public Health Service to investigate."

Val jumped to her feet. "I call for a vote."

Maggie rose slowly. "Ladies please, let's calm down. This is a serious issue; we all know that. And we all know that the town is divided on it. I, for one, believe that we, as a respected organization here in town, should have a voice in this."

"But Maggie," Val Brown shouted, "I didn't join the Women's Club for this. I call for a vote."

Trish Vanderville joined Maggie at the front of the room. She reveled in public controversy and knew how to fan the flames. For effect, she pushed her sweater up to her elbows, revealing peace symbols tattooed on each forearm.

Maggie smiled inwardly.

The rest of the room shuddered.

Trish wore a devilish smile. "It is interesting, Val, the main reason I joined the Women's Club was to make a difference on issues like this. I believe more harm is done by remaining silent on these issues than by discussing them openly. I support Polly, and firmly believe that if we hang together, we can make a difference."

Even the spiders in the rafters were aghast.

Without question, Trish was an excellent teacher. She had quickly earned the affection of her students and their parents. The old guard in the Women's Club, however, was used to getting their way. They had never been comfortable with Maggie, but could usually swing the vote. Now Trish, and Polly, and even Lucy had swung to the left.

After an awkward silence, Maggie brought the motion to a vote. It failed, four in favor, fifteen against. The meeting was quickly adjourned. There would be no Five Hundred today. The old guard filed out knowing things would never be the same. A fundamental taboo had been violated. You don't bite the hand that feeds you, certainly not in a company town. Sign a petition? Go public with your concerns? What could they be thinking?

Polly sat quietly beside Lucy and Trish as the room emptied. Storm nudged against her, ready for a run.

Maggie was beaming when she returned, "Hallelujah girls, the fat's on the fire!"

"Fifteen against, why so happy?" Polly sighed.

"Ha-ha, trust me, most of them will come around."

Trish paraded around the room chanting, "Free-da–people, free-da-people."

Maggie chuckled. "You know what Susan B. Anthony would say? Find a man to blame.

No, just kidding. We need to build an alliance."

"I am with you there, Sister." Trish beamed as she marched by with Storm at her heels.

Polly sighed. "Oh dear, dear, dear. Johnny's going to kill me."

Chapter 10

Catalytic Action

Johnny poured coffee from a thermos as he scanned the control console. It was five in the morning and he had things to do. A wall of gauges, levers and lights flashed before him. Two weeks ago, the daily target had been raised to one hundred and seventy tons per day up from a hundred and fifty. The converter had been acting up ever since. The conversion gauge now read ninety-six percent; the target was 99.5. A swing of two or three percent was serious, output dropped, more sulfur dioxide poured out of the stacks, and less acid was produced. Johnny pulled a technical manual from the shelf entitled: *Contact Process for the Manufacture of Sulfuric Acid*. He opened the manual to the first page and read: Iron pyrite is burned in the reactor and produces sulfur dioxide gas. *After passing through the precipitator and the drying tower it moves on to the converter. Vanadium pentoxide, a catalyst, is used to convert sulfur dioxide gas into sulfur trioxide.*

$$SO_2 + O_2 + V_2O_5 = 2SO_3$$

He scratched his head and tried to say it, "Va-na-di-um pent-ox-ide. Sounds evil." He read on: V_2O_5, *a poisonous orange solid, is the most important compound of vanadium, and may be fatal if swallowed or inhaled. It causes irritation to skin, eyes, and respiratory tract. Unlike most metal oxides, it dissolves slightly in water to give a pale yellow, acidic solution.*

"I knew it was evil."

The Vanadium Corporation of America down in Uravan needed the Argentine's acid to enrich their uranium, but he didn't know that vanadium was used in making acid. How about that, right here on the Western Slope? He flipped to the chapter on maintenance.

The converter contains a series of 16' diameter mesh trays that are stacked one above the other. Each tray is filled with approximately 10" of vanadium pentoxide pellets. As sulfur dioxide is blown up through the trays it is converted to sulfur trioxide. The pellets need to be regularly replaced based on output levels (see table below). Caution: If converter is run at excessively high levels, the pellets will cake up. This decreases the conversion rate and dangerously high levels of sulfur dioxide are

Rico Acid Plant 1960s

lost out the stakes. Standard respirators cannot filter out high levels of sulfur dioxide.

Johnny stepped out for a smoke at six-thirty. He walked up above the plant to the St. Louis Tunnel. Dawn was breaking and the skyline had a burnt orange glow. He leaned against an empty ore car and enjoyed the play of light on the mountains.

When he turned back to the plant, a column of dirty fumes poured out the stack, a second later the alarm bell started to blare. He crushed out his cigarette and rushed inside. The converter now read ninety-three percent. As the alarm blared on, the cords in his neck tightened. Setting protocol aside, he went into shutdown mode, closing valves, throwing levers, and thankfully, cutting off the alarm. He raced off for the converter, but caught himself at the door. For some strange reason the image of a lizard came to mind. He thought of what Polly had said, "Sometimes, you need to move forward, and sometimes you need to step back, to move forward."

He took a deep breath, exhaled slowly, and returned to the desk. Ignoring the phone, he quickly sketched out a plan:

- ✓ cool reactor slowly
- ✓ gloves, goggles and respirators for all crew
- ✓ bring vanadium pentoxide from storage
- ✓ test the rocker
- ✓ brief crew

He finally walked over to the Western Electric wall phone and picked up the receiver,

"This is Johnny."

"You in shutdown mode?" screamed Chester.

"I am."

"On whose authority?"

"Mine. The converter dropped to ninety-three."

"So what? It'll cycle on you. The wind is blowing up canyon, no one gets hurt."

"The reactor's gonna blow."

"Hang on, I'll be there in a minute," Chester hung up.

Hang on bullshit, Johnny thought. He'd crossed the line, yet again, and dammed if it didn't feel good.

Roy and the rest of the crew stood at the doorway taking it in. Roy pulled Johnny aside, "You sure you want to do this?"

Johnny winked and stepped back, "Ok guys, gotta go, gotta go, get your safety gear and get down to the converter pronto. Roy, you get the pickup and pull up two drums of the orange pellets from storage, bring 'um down to the converter, go, go, go."

Minutes later they all stood in front of the converter. Johnny checked their gear and briefed them. "Listen up, boys. I am only going to say this once. The converter's been acting up for the last couple of weeks, so we need to look inside, see if the catalyst is caking."

He grabbed a handful of the catalyst from the drum, "This here is called vanadium pentoxide. It's nasty shit. Don't be rubbing your eyes or scratching your balls. It doesn't work when it cakes. That's what's killing our conversion rate and stinking up the town." His eyes narrowed and he slowed down, "We're going to open her up, pull out the mesh trays, and rake out the cake, right? Then, we'll screen the rest through the rocker, right? And finally, we'll top off the trays with new catalyst, right? One, two, three, just that simple." Again he paused and looked around, "Any questions? Good, Roy and I'll go first, Billy and Slim your next. Let's work it."

Johnny slowly cracked the seal on the manhole like he was opening a hot beer. Just as Chester drove up, a blast of yellow smoke spewed out, hissing and steaming, "Come here, Carnifax."

Johnny locked her down and walked over to his boss.

"Who the hell are you, calling a shutdown?"

"Gotta a big problem here."

"I'll decide if we gotta big problem. The converter cycles up and down, no cause for alarm."

Tell that to the alarm bell, Johnny thought, but then held his tongue, "I've checked the shift log and it's been dropping steadily for two weeks, since…"

"Since, what?"

"Since, we started crowding the system."

"You still don't get it, do you Carnifax? You gotta problem, you go through me."

"I ain't arguing, but the whole system is backing up."

"Look, shit-stick, I don't care if the sky is falling, you call me. I call a shutdown."

The crew looked on in silence. They were stuck in mid-stream, not sure which way to jump. All but Roy, he'd gone on red-alert. He ambled up to the truck, cracking his knuckles and loosening up his arms. Johnny saw him coming and shook him off like a pitcher wanting another pitch. He turned back to Chester, "You want me to call this off?"

The wind had shifted and the stack smoke was drifting toward town. Chester fiddled with his keys. His face had gone red and puffy, "Just get her done and be down at the Burley Building first thing, tomorrow!"

As Chester drove off, Roy nudged Johnny, "You want me to handle this?"

"Maybe, let's see how it goes tomorrow."

They walked back to the converter, "Ok boys, watch me and Roy. Then, we'll switch off. We get this cleaned up and we can start making acid, and what the hell, maybe we'll get to keep our jobs."

The crew sensed that Johnny had won that round and quickly fell in line. After ten minutes inside the converter, Johnny and Roy crawled out, filthy and exhausted. They stumbled upwind, yanked off their respirators and sucked in the clean mountain air. Fresh and clean, the next team crawled in. By two, they'd finished and were back online.

Acid Plant Control Room, Edgar Branson, left, Jack Darnell, right

Roy collared Johnny as they headed for the control room, "You did good there, Mr. Foreman."

"We'll see tomorrow."

"Say Johnny, what's a shit-stick?"

"Can't say, gotta ring to it though, don't it?"

"Sure does." Roy smiled changing the subject to his domestic escapades with Ronda. "I want to tell you, the singing worked. Took about twenty minutes. Cold as hell outside. The dogs are howling, lights going on in the neighborhood, and sure enough, the door swings open, and Ronda hauls my bony ass inside. That's what she calls it, bony."

"Stick with me, son. Look, I want to recheck the shift log. I'll join you later at Lucy's."

"I'll be there. I am singing tonight."

In the Air Force, Johnny logged hundreds of hours in a B29. The control room was like the cockpit. He paged through the shift log. The conversion rate was fine until they started crowding the system. No cycling, as Chester claimed. He checked the conversion indicator on the console; it was back up to ninety-nine, in range. Sure, there were problems, but none that couldn't be fixed. For the first time, he felt a measure of control. He had handled the crew pretty well; everyone had pitched in. Yeah, this was a good day. Polly would be pleased.

Chapter 11

REALIGNING THE STARS

Johnny felt great as he walked into Lucy's that evening. There wasn't an empty seat in the house. Roy and Ronda stood nose to nose on the dance floor, bickering. Johnny waved. She damn sure was a wide-load.

Roy rushed over. "Not a good day for the girls. Polly flipped on ya and Ronda's back on sugar donuts."

"What'd Polly do?"

"She spoke up at The Women's Club, complained about the plant. I guess the big boys' wives were there. She really stirred the shit this time, town's in an uproar."

Johnny got the full story and then headed home. Their first year of marriage had been rough, two pit bulls, dueling it out. They almost split, but worked their way through it, came up with their own set of do's and don'ts. They learned to talk things through, especially the important stuff. He was no angel, but he was trying to make it work. So what in hell was she up to?

Polly was at the stove when Johnny arrived. She checked the biscuits in the oven and joined him on the porch, smiling sheepishly, "Hi honey."

Johnny pulled off his boots, "Had a busy day, did we?"

Polly cringed, "I didn't know they'd be there."

"Yeah well, I didn't know you'd be there."

They stood motionless, suspended over a void. She burned with guilt. He steamed with anger. A false step would take them down. He wanted to lash out, but deep down he sensed, they were both struggling with the same demon. She was the best thing ever. Sure, he had a crazy streak, played it too close to the edge. But together, that was special, and the baby coming. He looked down at the new crib and exhaled. Hell, it's only a job, easy come, easy go. Maybe they'd move down to Moab and prospect for uranium. He pulled her in and kissed her on the forehead.

"Come on girl, them biscuits smell great."

Polly melted. In a flash they were safe, off the tight wire, across the void, together.

"Johnny, I know I was wrong. I shouldn't have made that speech."

"Let's eat darling, we've got a lot to talk about."

After dinner, they sat out on the porch. A glittering mantle of stars stretched across the sky. Johnny sipped his coffee, "So darling, tell me about your day."

"You know, we've talked about the pollution, and I am convinced they drain the ponds. I went to the library and found the newspapers full of stories about the plant, and I talked to Maggie Fagan, and—"

"Slow down, Dear."

Johnny rocked back in his chair and tried to sort things. His anger smoldered just below the surface. It wasn't Polly. They were on the same side, just working at cross-purposes, nothing new there. Had he been wrong shutting down the converter? Could the problems with the reactor, and the crusher, and the tailing ponds really be fixed? Would they listen? This thing was far from over.

Polly started over, "I know we should have talked it over, but I never dreamed that the high and mighty, Mrs. Elizabeth Pritchard, would be there. I got a little angry when I saw the dead fish and then I started reading up on the problems at the library, and I got real angry. Take a look."

Johnny read aloud from her notes, "Explosions, shutdowns, sulfur dioxide choking off the town." He turned the page. "Numerous complaints at town hall meetings, lawyers' letters."

"What's this in quotes? "every available precaution has been taken." Is that what they said?"

"It's all public record. That's what set me off. I showed it to Maggie, and well, you know the rest."

Johnny picked up the phone on the third ring.

"Hello, yeah Roy, what do ya mean she left you? Okay, okay, hang on."

Johnny covered the phone and turned to Polly, "Ronda's gone, hitchhiked to Las Vegas, going to work in a casino, hostess or something."

Polly shook her head, "You and your trashy friends."

"Yeah Roy. I can hear you. Tanya? Who's Tanya? What? New girl from Stoner? Who's Leon? Hang on."

"Roy got into a fight with Leon, a cowboy from Stoner."

"And Tanya?"

"She's Roy's new girl. Roy was singing Hound Dog and Leon starts insulting him, howling like a hound dog. Roy knocks him cold. Tanya's got an Elvis thing and now she's moving in with Roy."

"And Leon?"

"Turns out, Leon is more of a Hank William's fan, don't care much for Elvis, says it drove them apart. He and Roy are best buddies now."

"Sorry Roy, got a lot going on here. Ok, listen, you get yourself home. We got a big day tomorrow. I'll be dammed. Is that what Leon said? Ok buddy, get to bed." He hung up and smiled.

"And what did Leon say?"

"Said, the acid plant is polluting the river. Said his cattle won't drink the water. Said he's gonna drive his herd up to Rico, stampede 'um through the acid plant. And Roy said, he was proud of you."

"Roy said that? Well, bless his hillbilly heart. Are we done here?"

"So Polly, we got to get back on the same track. I've been working at being a supervisor, and I still believe I can talk some sense into the big boys, but no more rabble-rousing speeches."

"So what's your plan?"

"Let's finish with you first. I've never been one to tell you what to do, but you've got to cool this Mother Jones crusade, at least for now."

"I hear you, but just to set the record straight," she tapped on her notebook, "there's not a thing in here that isn't a fact. It's common knowledge."

"Just cool it for now."

"I said I would."

"Oh, I must have missed that. So for my part, I've been looking around the plant. There's lots of things wrong, but most of it won't take much to fix and it all makes good business sense, if they'll listen to reason."

"And if they don't?"

"Then I reckon we're out of here. Head down to Moab. Buy us a Geiger counter, get filthy rich."

Polly jumped up and locked her hands on her hips, "No damn way, we're not going to Moab or anywhere else. For one, we can't afford to move, we're a month behind on our rent right now, and two, we're through running. We ran from West Virginia. It's time we planted a stake in the ground and stopped running."

"Wish me well. I've got a meeting with Chester at the Burley Building in the morning. I'm gonna make my case."

Chapter 12

WHERE THE HURLEYBURLY'S DONE

Johnny had been at it since four a.m. with Polly right by his side. The second pot of coffee perked on the stove. He had written and rewritten his notes and finally got it down to three points on one sheet of paper: fire prevention, maintenance, and tailings disposal. These things had to be fixed. The cost could quickly be recouped with the output gains, which he felt was a strong selling point. Not that the Argentine seemed to care, these corrections would also greatly reduce the pollution.

Polly typed it up on her battered Underwood, "It's a good plan."

"I am damn well ready to talk, if they're ready to listen."

"Just stay cool."

Johnny knotted the laces on his boots for the third time, "Yeah, that's what I'll do."

"Right," Polly rolled her eyes.

"Hold onto the carbon copy, darling."

Johnny was a gifted shade tree mechanic, a born hot rodder. He was cool in dealing with technical problems, but not cool with business problems, especially those involving lawlessness. There he got trapped between scheming and fussing. Scheming came from his father: a mine supervisor, a moonshiner, and a gifted, horse trader.

Johnny Senior always said, "When it comes to getting along with your boss, try to figure out how you can help them get what they want. If you can do that, you've got a lot better chance of getting what you want."

Surely that would work with Chester. Just show him how to increase production and control costs, that's what he wanted, and all else would follow. Yet, he was troubled. There was nothing in his plan that Chester hadn't thought of, yet, he'd done nothing about it.

If scheming didn't work, Johnny went to fussing. Anger and injustice pushed him in that direction. Fussing was the exercise of raw power, hurting the bad guys or taking their stuff. You gained an edge with bullies, thugs and crooks in the coalfields, if they knew you could hurt them. He'd recently added lawyers and accountants to the list, the educated outlaws. Threats, intimidation, and aggression were all fair game in fussing. As a child, he believed in the moral high ground, just doing the right things. Lately, he found himself doing the wrong things, in order to do the right things, and he done some jail time for it. He wasn't proud of this, but it seemed to work, sort of. He carefully folded his plan, stuck it in his pocket.

Polly kissed him at the door, "Stay cool Johnny, staaaaay coooool."

"And you, staaaaaay off your soapbox," he grinned and headed for the Burley Building.

He walked down Campbell Street, crossed Glasgow Avenue and turned right. The sky was clear and the streets were plowed, always a good beginning. He walked past the firehouse and stopped in front of the Burley Building, an impressive stone structure on a 50' x by 80' foot lot. J.W. Burley & Co. opened their wholesale and retail house in the spring of 1890. A grocery store now occupied the left side of the building and a drug store on the right. Between them ran a stairway up to the head office of the Rico Argentine Mining Company.

"Hello there."

It was Maggie coming up the street.

"Well, good morning."

"You're early, Chester won't be in 'til eight. Com' on up." She unlocked the door at the top of the stairs. "You can wait in there. I'll make some coffee." She disappeared into a tiny kitchen. Johnny took a seat in a small office that looked out on the eastside of town, a waiting room of sorts. He picked up a copy of the *Dolores Star* and flipped to the sports page. Tonight, the Dolores Miners played their archrival, the Dove Creek Bulldogs, flatlanders to the local fans.

Minutes later, Maggie returned with coffee, "Here ya go Johnny-Reb."

"Where's that Johnny-Reb coming from?"

"Brad and I love our Civil War history and we've been reading about the Carnifax Ferry and old John D. Carnifax."

"That was my great grandfather's brother. He was damn sure a rebel."

"You got quite a noble bloodline."

"Thanks, there's no substitute for good breeding. It says here that young Brady Fagan is quite the basketball player."

"We're real proud of him. The regionals are tonight, winner goes to the state."

"It says he made twenty-five points in the first half of the Telluride game."

"That he did. He got kicked out in the second half. Telluride had this gorilla at center; he couldn't make a layup, but sure could throw those elbows. He chipped Brady's tooth in the first half. They had a hometown ref to boot."

"What happened in the second half?"

"Right off, Brady takes one on the nose, starts bleeding; got Kleenex stuffed up his nose. Then he throws one back. Ref calls a technical. Brady argues and he's thrown out."

"They won big, though."

"Funny how things go," Maggie smiled with a twinkle in her eye.

At eight o'clock sharp, the front door slammed. Chester stormed down the hallway, stopped at the door, and scowled, "I'll see you in thirty minutes."

Maggie turned to leave, "Gotta go."

Johnny patted her on the shoulder. "Tell Brady to stay in the game tonight."

Maggie stepped up close to Johnny and grabbed a button on his coat, "The world is full of gorillas and hometown refs. You stay in the game this morning."

Johnny looked up as five, well-fed men scurried down the hall, a blur of dark suits, white collars, and shined shoes. Maggie sent them into Stanley's office.

"He's on the way gentlemen; please have a seat in his office."

She stuck her head in the doorway, "The Emperor Penguins have arrived."

Johnny whispered, "Who are they?"

"The big boys from Salt Lake and Cortez, lawyers and bean counters."

Maggie ran to catch the phone, "Hello, oh hello Lucy darling."

"I just had a Stanley sighting. He's turning onto Glasgow from Soda."

"What's he wearing?"

"I can't tell yet, high boots and riding britches. He's got them hitched up to his arm pits, the ones with the high pockets."

"High pockets, oh God Lucy that is funny. What else?"

"A top coat with a fur collar."

"How about the hat?"

"Hum, not the fedora, wide brim, can't tell yet, and a bandana around his neck."

"Big game hunter, you think?"

"Not sure. It's pinned up on one side."

"An Australian bush hat?"

"That's good, an Australian General maybe?"

"No, hang on," Maggie cackled, "I've got it, it's Teddy, Teddy Roosevelt leading the Rough Riders."

"Damn girl, you nailed it. And Maggie, a funny thing happened last night. The bar was buzzing about the donnybrook at the Women's Club and then Slim Webb got up a betting pool on when Johnny would be fired."

"Spineless pond scum."

"But then, the mood changed, people started opening up about the acid plant. They decided that if Johnny gets fired, he and Polly get the pot for travel money.'

"How about that? Sorry, gotta go, High Pockets is coming up the stairs."

Stanley strutted down the hall tapping his cane on the floor. He glanced in at Johnny and kept moving.

Stay cool, Johnny kept repeating. It was eight-thirty. He'd been there for an hour, a waiting game perhaps, okay, I can play. He looked up at a large, sepia photograph that hung next to the window. The words: Pro Patria Mine, 1912 were engraved on a small brass plate. An aerial tram suspended from huge wooden towers cut diagonally across the picture. It ran east from the mill below town, up Mantz Street and then climbed several thousand feet to the mine. The tram house up at the mine was partially blocked by the tailings, a towering pyramid of gray rubble. In its heyday, tons of lead and zinc ore were loaded into tram buckets and hauled down to the mill. There it was processed, loaded into gondolas, and shipped to the Durango Smelter on the Rio Grande and Southern line. Johnny marveled at the engineering involved in it all, then glanced out the window. It took a minute or two for him to realize that the view out the window was the same as the photograph, only fifty years later. Both looked up Mantz Street to the mine. Much had changed. Yet the tailings remained, joined by dozens more that scarred the mountain like a battlefield.

Johnny looked back and forth at the photograph and out the window, the past and the present. The snow chutes high on Blackhawk Mountain had been there since the ice age. They were in the photograph and they were up on the mountain. He rubbed his chin. He wondered what it would look like in another fifty years.

Aerial Tram from Pro Patria Mine to Mill, 1912

Maggie popped in at nine. "Chester's ready, last office on the left."

Johnny knocked on the door and stepped inside. Chester sat behind a battered metal desk stacked with file folders. Ore samples were on every shelf. He quietly flipped through an open file, not bothering to look up.

Stay cool boy, staaaay cooool, Johnny counseled himself.

The Argentine owned over three thousand acres of claims in Rico with dozens of mines that opened and shut based on the whims of the market. Chester knew every tunnel and shaft on the mountain.

He slowly closed the file and looked up. No chair was offered. Chester had mastered the art of scheming.

"I just called Paul Spitzer, your brother in law; he says you were in the Air Force in Korea. What'd you do?"

"Ordinance specialist, Tactical Air Command."

"How come it's not in your file?"

"'Cause it's top secret and it's none of your damn business." Johnny growled and then caught himself.

"Ordinance specialist. What the hell's so secret about that?" Chester barked.

"We moved nuclear weapons in and out of Seoul on a regular basis, weapons grade uranium. You've heard of that, haven't you?" Johnny saw by his reaction that he was impressed.

Chester moved around his desk and stood directly in front of Johnny, "Well. Mr. Tough Guy, you're on a short rope here, anymore complaints and you're gone."

Johnny bit his lip and paused. He noted a minute shift in Chester's eyes, a slightly exaggerated tone. He had softened. A subtle shift of power had occurred, a glimmer of hope.

Hat in hand, Johnny remained standing, "Yes, that's clear."

"You gotta problem, call me first."

Johnny quickly considered his next step, "Yes sir, but can I ask you a question?"

"Make it short."

"Say I am driving by your house and it's on fire, and you're down at Stoner, you want me to call you first, or call the fire department?"

Chester glared at him for the longest time then slapped him on the back, "Get your sorry ass out of here."

And so it went for several minutes, two dogs circling each other, barking and growling, testing boundaries, strutting their stuff, secretly enjoying the banter.

Johnny senses he had a slight opening, "You know, there are ways we can be a lot more productive up there."

Chester glanced at his watch. "Make it quick."

Johnny pulled out his plan and laid it on the desk. "You've got three problems: fire, maintenance, and tailings. Get them right and the gains will cover the cost in no time"

"Leave it with me. Fire you say?"

"Take a look. You got pyrite dust caked on everything. Piss poor ventilation and the jaw crushers throwing sparks."

Maggie stuck her head in the door. "Chester, Stanley wants you in there now."

"Ok, Carnifax, you're on probation. And tell that pretty wife of yours to stop her rabble rousing."

Johnny stopped at Maggie's office on the way out.

She smiled, "You still on the payroll?"

Johnny nodded, "I am on probation."

"Me too."

"You and Polly gotta cool it for a while. I think we got a chance here."

Chapter 13

Texas Hold 'em

"There was a corporate philosophy (at British Petroleum) that it was cheaper to operate to failure and then deal with the problem later, than do preventive maintenance." —Scott West, the EPA special agent in charge of the investigation of the 2006 Alaskan Pipeline Spill.

The quarterly review was held in the boardroom of the Burley Building, the sanctum sanctorum of the Argentine Mining Company, replete with mahogany furniture, Persian rugs, and a silver tea set. Stanley T. Pritchard stood beside an overhead screen wielding his cane as if he was leading the light brigade. Corporate lawyers and accountants sat around the boardroom table in hushed attention. Stanley tapped at the first of several abbreviations on the screen, pivoted, and thrust his cane at Bill Parma, one of the corporate lawyers from Cortez.

"As you all know, VCA wants us to double our output. If we can't, they threaten to go with another supplier," he whispered cryptically, "so Bill, what can we do?"

A sinister smile flashed across Bill's face, "Tie 'em up in court, sue for breach of contract, whatever."

"Good. Get a short list of other potential customers in the region. Our shipping costs will be higher.

He thrust the cane at Rob Tighess, the Corporate Financial Officer from Salt Lake City, and a polished sycophant, "Stan, run the numbers for him."

"Yes sir, and I've got the year-end financials. You won't be disappointed."

"Later." Stanley's cane swung to Dan Miller, the other Cortez lawyer. "Dan, you need to talk with this Meyer guy, the Dolores mayor. He's on the phone every day complaining about the river."

Dan flipped through his file. "What do you think? The mea culpa letter or the human error letter?"

"No, just take him to lunch; remind him of our contribution to the local economy. He plans to run for county commissioner, we could help. Just tell him no more petitions."

Dan gathered up his files. "I'll call him today, set up lunch. Is that it?"

"Hang on; you need to put a letter together to the Rico Town Board. I am tired of the frozen pipes. Our taxes support this burg. They can damn well pay for the new water line without our help. Get a letter out before the town board meeting next week."

"That's it for the local issues," Stanley turned off the overhead, "thank you gentlemen."

Bill and Dan packed up their papers and files.

As the door swung open, Chester stuck his head inside, "You ready for me?"

Stanley shook him off and closed the door.

Chester walked slowly back to his office shaking his head. What a prick. He picked up Johnny's plan. Scrappy kid, Johnny, kind of liked him, but I may have to let him go. He was right about the tailings, gotta dry stack 'um. Maintenance, now that's different. Chester had never won that argument. He could hear Stanley now, "Cutting corners on maintenance, is how we survive up here. We have to learn to live with more risk until the market turns."

And yes, fires were a concern.

More and more, Chester was excluded from these meetings. He had never clicked with the corporate guys. Theirs was a world of leverage and loopholes. They had no clue when it came to operational issues.

Chester scooped a gob of Dr. Hess Udder Ointment from a can in a desk drawer and worked it into the cracks and splits of his hands. They were massive bear paws; gnarled and calloused, splitting at the tips and joints, miner's hands. He felt trapped between generations. The early miners were a rugged breed, but innocent in their enthusiasm. The chance to stake a claim, to get rich, to blow it all in a poker game, and then to start over—that was the good life. They were fascinated by the wilderness, yet easily seduced by the comforts of home: a hot meal, a cold beer, a soft bed, and good company.

Rico was luckier than most mining towns. It had been blessed with good leaders that built the business and the community. It was simple, really. Neither could grow without the other. David Swickheimer known as 'the backbone of Rico' triggered the first mining boom in 1887. He made and lost fortunes. The Rico State Bank, which he owned, went into receivership in the panic of 1907. He paid all his depositors, dollar for dollar including six percent interest on their money because, because it was the right thing to do.

In the thirties and forties, Robert and Betty Pellet kept the Pell-Eyre Mining Company running and kept Rico from becoming another ghost town. Robert served as mayor for three terms and Betty was a long serving state representative. And C. T. Van Winkle, who was brought in by the Rico Argentine in 1937 was also good for the business and the community. He opened the Blaine Tunnel Mine and built a flotation mill on Silver Creek. And then, was mysteriously fired in 1948. They say he didn't play ball corporately. C.T. was succeeded by Stanley T. Pritchard, who closed the operation and reduced the crew from two hundred to fifty.

Chester worked the grease deep into every crack as he reread Johnny's plan. Together, corporate power and unbridled capitalism were a ruthless juggernaut with its own cold logic, he concluded.

Stanley huddled the boys from Salt Lake in the farthest corner of his office. "You know the drill, it's Texas hold 'em. We gotta stay in the game, but not raise the ante. Acid is the only thing we've got going right now, so we hold on. We don't fold, that's what the little guys do. We can raise the ante when the market comes back. For now, we trim the payroll; sweat the assets, and ignore the righteous."

They nodded gravely, willing co-conspirators.

Stanley turned to George Thornton, the Chief Legal Counsel for the Argentine.

"George, you got a buddy with the State Attorney General's office, right?

"Classmate in law school."

"As you've heard, our friends downstream have been circulating a petition to the State Bureaus of Health. See what you can do to get the PHS off our back."

"On what grounds?"

"Hell, George, you're the lawyer; we're in a Cold War. Invoke national security. That's what the Atomic Energy Commission does."

They shifted nervously as they heard a knock on the door.

Maggie stuck her head inside. "Sorry."

Stanley stepped to the door.

"It's Bill Meyer, the Dolores's Mayor. He's been calling all week."

"Did you tell him I was in?"

Maggie shook her head.

"Good… say…say, I am up at the plant. Tell him he'll get a call from Mr. Dan Miller today."

He waited until the door was closed and continued. "Ok, we're done here, except for Stan. Now boys, I don't want to see a paper trail on any of this."

Stanley and Rob Tighess refilled their coffee cups and stood by the window.

"I've got a little something for you, Stanley," Rob announced handing Stanley a gift wrapped package.

"Well isn't that lovely." He carefully removed the bow and unfolded the wrapping paper. "Oh my, my, chocolate pecan clusters, my favorites!"

"Just a little something for all the help you have given me."

"Thank you Rob. I feel we have special relationship."

Rob beamed, "I think you will be pleased with the numbers. Had to move some costs around and carry over some outstanding invoices, but it was a very good year for the acid plant."

They returned to the table to review the financial projections.

"Well, well, well." Stanley purred. "This looks great. What does the Chairman think?"

"Oh, he's delighted. Do you want my opinion?"

"Yes, of course."

"Another year like this and you'll be back in Salt Lake, running the whole Western operation. That's the scuttlebutt."

"Hmmm, well, well, well. Let's keep that quiet."

They shook hands at the door.

"And Rob, if there is anything I can do to help you out, anything at all, please ask."

Stanley walked down to Chester's office, "Sorry, that took so long. Come on down to boardroom."

Chester followed.

"Is Carnifax gone?"

Chester stood like a bull in front of Stanley's desk. He looked up slowly with a blank expression. "No."

"No?"

"No. He's a good man, hard to replace."

"Pull a Navajo from the mine."

"I would, if one was ready."

Stanley fidgeted with his cane.

Chester stared at his hands: soft, delicate, pink palmed, well clipped, paper-shuffling hands. Could they steady a drill, pound a sledge, brace a roof, ride a jackhammer?

"Need I remind you, that your job is to deal with these problems?"

"My job is to run three thousand acres of claims as well as that stinking plant."

"That stinking plant is paying the bills right now."

"And you're running it into the ground, dammed if that makes any sense."

They were stuck, hand cuffed in an unholy alliance. With Stanley, it was all about vanity, ambition, and power. He wanted to leave, one more good year, get that promotion and head back to Salt Lake. He'd fire Chester in a heartbeat if he had a replacement.

With Chester, it was more about survival. He wanted to stay, oh he was fed up with wonder boy, but he was settled, his kids were in school and he had his own claims to work.

Stanley modulated his tone. "Just keep her running for now. If zinc goes up, we can put some money back into her." He paused, then swung his cane wildly, "And tell Carnifax to put a muzzle on that wife of his."

Chapter 14

BITAH HONEEZGAI (SICK ALL OVER)

Several Weeks Later

When Norman was too sick to work, he moved up to Rico with Chee. Below the plant, they built a traditional Navajo *hooghan*, out of ponderosa pines they'd cut near Stoner. It had six sides and a doomed dirt roof with a single smoke hole in the middle. The walls were made of stacked logs, notched together at the end. Once the walls were up, they packed them with adobe mud. The eastern facing wall had an open doorway to let in the rising sun. Inside there were no windows or internal divisions, just one large room with a circular dirt floor. A cook stove stood in the middle, the stove pipe extended out the smoke hole. When it was done, the medicine man came up from the reservation to bless it.

The *hooghan* is central to the Navajo creation story. The "Blessing Way" song tells how the first *hooghan* was built by Coyote and Beaver, for First Man, First Woman, and First Talking God, and Second Talking God. Beaver gathered the wood and told Coyote to sit on the ground and face east. The *hooghan* was build around Coyote. The dome roof symbolized the sky, the floor symbolized the earth. The *hooghan* creates a harmony between man and the mysterious forces of life.

As dawn broke that morning, the sun spirit crept in the door. Norman sat near the stove chanting in a shaft of smoke and light, his soul shackled to the earth. His face, a granite slab, furrowed by the joy and sorrow of a life well lived, a fading warrior. And then he was off, floating free, riding the Holy Wind, feeling the *bii'asti*.

Out on the reservation, a gypsy wind rolled down from the Lukachukai Mountains, ducking in and out of mines and dog holes. As a boy, Norman had drilled and mucked in the toxic darkness. Moaning low and braided with yellow fumes, the wind gusted north into Colorado, running through the Slickrock Mine. It was here, at fourteen that Norman learned to drive the great jackhammer. Up above, a thunderhead churned, white and whirling, cresting like a wave. It sent a twister on a run, cutting across the sandstone mesa. Funneling up the Dolores River, it triggered a lightning storm. *Leetso*, a winged serpent, spewing yellow venom, sprung from the fire storm. A burning tremor shot across Norman's chest and tightened around his throat. Then *Rado*, a luminous red elk, trotted into the acid plant. A brown plume drifted from the stacks and wound its way around the *hooghan*. Norman's head snapped back. He convulsed in pain, his lungs on fire, his throat in spasms, struggling to cast out the bad wind.

He fought back, cleared his throat, and drifted back into a dream shadow. There were fewer good dreams these days. Maria, his Lakota wife, had died of tuberculosis ten years ago. On a good day, they flew as one over the high peaks in the clear air and the Holy Wind. He dreamed of Chee running with the wolves as a child. This too was good. But more and more, the bad wind had come, unleashing *Leetso* and *Rado* to spread their venom. Or, were these dream as well, ghosts of dead miners, returned as whirlwinds and lightning? Was it they who lit the fire in his chest? Was it they who stole the wind from his lungs? He pushed the unburned end of a log further into the stove. What had gone wrong?

In December, he had gone to the reservation. A diagnostician, a crystal gazer, told him that he needed the Navajo Wind Way ceremony to help cure his breathing problems. A medicine man performed the ceremony that involved five nights of purification, prayer service, sand painting, and sweat baths. He sought to drive the bad wind from Norman's body and restore harmony between his spirit and the spirit of nature. Norman was at peace for several days and then the fire returned. Weeks later, Chee drove him to the reservation hospital in Shiprock. There he took the white man's medicine. And still the fire burned inside him.

Even with his sickness, he had been a skilled hunter. Last year, he killed an eight hundred pound elk on Blackhawk Mountain. Nowadays, he chanted and chopped wood in the morning. At midday, he walked into town, collecting his mail and visiting with friends. Chee wouldn't let him go into Lucy's anymore. He'd get to drinking and wouldn't come home. Lately, he had taken to buying a pint of whiskey on his rounds and drinking at home.

After his trip to town, he hobbled up the hill to the sweat house that he and Chee had put up to the north above the *hooghan*. They dug a shallow pit for the floor and covered it with juniper bark and branches. Then built a frame over the pit with three forked juniper logs and tied them off on top. There was no smoke hole in the house. The door faced east with a blanket drawn over it. The rest was thickly covered with earth.

Once outside the sweat house, Norman threw himself into splitting wood. This he could do and it kept his mind from drifting. The visions faded. Yet, always the stealthy predator, the acid wind waited patiently, and then drifted through the trees. The fire in his lungs returned. He doubled-over in a fit of coughing and stumbled to higher ground.

―⁂―

It was Saturday, but Johnny had to work. Polly and Dorothy Spitzer had driven down to Cortez to shop and take the kids to a movie and wouldn't be home until late. Johnny got home by six, washed up and had a dinner of leftovers. As he sat on the front porch smoking, he thought about his meeting with Chester. It wasn't all bad. I still have a job and maybe we can turn around the plant. As he looked up the hillside, he noticed a fire in front of Norman's sweat house. Better get up there. He'd been working a double shift for the last week and hadn't had a free moment. Two weeks earlier, he had promised to bring Norman some elk meat. He pulled a package of steaks from the freezer and headed up the hill.

Norman was pulling red hot lava rocks from the fire as Johnny walked up. The burning juniper sweetened the air. He watched silently as Norman carefully slipped a pitchfork under each rock and hefted it into the sweat house. Norman was a big man like Chee, but rail thin these days. He squinted in Johnny's direction trying to focus, "Is that you Crazy Bull?"

"Yes, it's me. I brought some elk."

"Ah, that is good. You have a good heart. Come sit with me"

They sat together on a boulder. A faint glow set off the skyline from the growing darkness. It was that fleeting moment when night and day came together, when hope and despair are joined in a fragile balance. Norman passed Johnny what was left of the whiskey. They sat quietly until it was dark.

"How are things at the plant?"

Johnny took a sip and passed it back, "It's bad, but we can fix it."

"It is a bad place all right. It poisons the air and the water." As he raised his voice, he was racked by another fit of coughing. He gasped for breathe, "I think, it angers the Holy Wind."

"And you Norman, how are you?"

"Not good. The white medicine doesn't help. I don't know why the fire is in my chest."

They finished the whiskey. After another long silence, Norman put his arm around Johnny's shoulder.

"Come Johnny, come inside."

Johnny was uncertain. Norman stripped and crawled inside. Johnny felt a blast of heat across his face. By now Norman was mellow, the whiskey had numbed the pain, "Come, we'll talk with the spirits."

Johnny stripped and crawled inside. Norman sat crossed-legged beside the glowing rocks. He put herbs in a water jug, blessed the water, and poured it on the rocks. A wave of steam filled the house. The heat was intense and cleansing.

"When the steam rises, the spirits are breathing," Norman whispered, then he started to chant, first in a low rasping murmur, then in a screeching falsetto.

Johnny was amazed, very strange, yet fascinating. He could only make out a word or two of English mixed in with the Navajo, "Holy Wind...speak...show me the way?"

Norman continued to chant, but softer now, back to a murmur. He was out there again, floating in the clouds. Sweat poured out of Johnny, as did the stress he'd been carrying since he came to town. He felt high, not a rowdy, whiskey high, more gentle, more peaceful.

Then Norman spoke directly to the Holy Wind, "Yes, I see the two fires...they burn brightly...but what...what do they mean?"

He turned to Johnny, "Do you see them?"

Johnny stared into the darkness and shook his head.

"The Holy Wind spoke…she's angry…but the vision…the fires … what is the meaning?" He began to chant again, but more forcefully, then he was up, dancing wildly. Sweat flew from his body.

Johnny wanted to speak, but didn't want to break the spell.

Norman continued, again and again, chanting, dancing, seeking a return of the vision. Then he turned to Johnny, "She's gone…we must find the meaning…we must solve our problems."

Johnny wiped the sweat from his eyes, "Tell me about the fires."

Norman pulled the bandana from his head and let his hair fall freely, then closed his eyes and began shaking his head, trying to separate the meaning from the mystery. And then he said,

"It was snowing…there were two fires in the distance. We must keep them burning, one on the mountainside, one in the meadow. Then the vision faded, the fires were gone."

"Did they go out?" Johnny asked.

"They were gone, that is all. The Holy Ones speaks in riddles, they only point the way." Then Norman drifted off, lost in his pain and his dreams, flying in the clouds over Lizard Head Pass.

Johnny left quietly. It was cold outside and he was buck naked. He fumbled around in the snow trying to pull on his long johns. He must have lost ten pounds. Then he started to chuckle. It had been a wild night. He didn't know what to make of the spirit world and Norman's vision. He felt like he knew more than he could explain. The Holy Wind had thrown pieces of a puzzle before them like turquoise stones on a trading blanket. It was up to them to fit the pieces together. Norman damn sure wasn't getting any better.

The ore house the morning after the fire.

Chapter 15

FIRE IN THE ORE HOUSE

Arm-in-arm, Johnny and Polly chatted and window shopped their way down Glasgow Avenue. It was eight at night and they were headed for Lucy's to meet the Spitzers for dinner. The moon was full and the sky had a luminous glow. It was Saturday night and Johnny still had a job. Polly cut a fine figure in a flower print, button-up dress and Johnny wore a pressed shirt normally reserved for weddings or funerals. It had been a month since he'd been put on probation. They'd gone through it all, every note and nuance. They were relieved, hopeful, yet guarded. Polly had agreed to step down from her bully pulpit. Matters at the plant hadn't got worse, but they hadn't got much better. Action had been taken on a few of Johnny's suggestions, nothing major.

Lucy's was buzzing. All heads turned as they made their way to the table where the Spitzers were seated. Some smiled, some frowned. The bar, like the town, stood divided.

Paul shook Johnny's hand; their eyes met knowingly, "You didn't expect a brass band, did ya?"

"No comment, your Honor." Johnny turned to the waitress," I'll take a Coors, what about you, darling?"

"Just a coke, thanks."

Polly sat down and silently watched Dorothy who was lost in some sort of card game. Laid out before her, were a stack of cards made from beer mats with words written on both sides. She nervously flipped them over, rearranged them in different sequences, mumbling to herself, and searched for mystical insights. A true believer in the power of symbols and words, she was a serious student of divination, having created her own runic alphabet and various tarot card games.

Polly waited for several minutes before she spoke, "Good evening Dorothy."

Dorothy looked up and blushed, "Oh, shame on me. Hello, Polly dearest."

"What are you up to?"

"Divination my dear, foretelling the future. The mysteries of life lie in plain sight if we pay attention." She picked up a card with Rico written on one side. "You see the word Rico," she flipped it over, rich was written on the other side. "Ha! Ha! Rico means rich in Spanish that sounds promising." She picked up another card with Dolores written on it, and flipped it over. "But, Dolores in Spanish means sorrow, Virgin Mary of the sorrows."

Polly nodded politely. Dorothy was wildly eccentric, but often right in her predictions.

"Polly you are so patient with me. Let me show you one more card."

She pulled another card from the deck and put it on the table. It had iron pyrite written on it. She flipped it over. "Iron pyrite is called fool's gold, thought to be a healing stone, a shiny, brassy yellow mineral. But Polly, in Greek, pyrite means "of fire" or "in fire." Pyrite was used to create sparks when struck, to start fires."

Johnny turned to Polly and arched his brows, the sign to move on.

"So Dorothy, what do you see in the future? What does this all mean?" Polly asked.

Dorothy set the three cards together on the table. She knew she'd gone on too long, "Briefly, and you understand there are many ways to interpret these cards. But briefly, Dolores, or sorrow cancels Rico or rich. There will be sorrow in getting rich in Rico. I feel it deeply. And then you add fool's gold and fire, it scares me. I don't understand it yet, but it is not a good sign. Maybe fools start fires."

"You are very good Dorothy, thank you," Polly said.

"So, how are you feeling? When is the baby due?"

"I am fine, the babies due in September. We can't wait."

Johnny paid Lucy for the drinks and turned to Polly, "What was Dorothy saying?"

Polly whispered, "You know, divination, fortune telling. Something about fool's gold, or fools starting fires."

The band tuned up in the corner. Roy gave Johnny a thumbs-up as he tested the microphone.

Polly looked up, "Oh Lord, is Roy on tonight?"

"I thought I'd surprise ya," Johnny laughed. "Roy and Tanya are working on a duet."

After dinner, Lucy appeared, with a mischievous grin and a cigar box under her arm.

"It's been a month now and you're still here."

"Officially, I am still on probation."

She set the cigar box down. "I guess this means you're staying, but that messes up our pool. What do you want to do with the jackpot?"

Somebody at the next table yelled, "Yeah Johnny, you're supposed to get fired!"

Polly put her hand on the box. "Let's give it to the Women's Club. Swings for the park."

Lucy pumped her arms in the air and shouted. "Swings for the park!"

A few clapped, then a few more, the momentum slowly spread down the bar. "Swings for the park," they chanted.

The wail of the fire alarm jolted the crowd like electric shock, no one moved. The volunteers were the first to move, running for the firehouse. Lucy's emptied quickly. To the north, the sky turned amber. Flames raged above the rooftops of town. The alarm continued to wail, echoing down the canyon. You could smell the stench of sulfur before you saw the smoke. Somebody yelled, "Fire at the plant!" and it echoed through the crowd.

Dorothy turned to Polly, "What did I say about fire."

Johnny looked at Dorothy and shook his head. She damn sure sees things coming.

He and Roy were the last ones out of Lucy's. When they heard the alarm, they unconsciously checked their watches and froze. As boys, they had heard the same alarm every day. It meant one of two things: it was either twelve noon or somebody's daddy had been hurt or killed in the mine. If it wasn't noon when the alarm sounded, every family in the coal camp ran to their porch. First the ambulance raced by. Every eye followed the next car. It was the mine supervisor coming to notify the next of kin. This was part of the baggage that both boys hauled out from Quinwood.

The alarm continued to ring. The Big Job Pumper raced by with volunteers hanging on both sides. It was a Ford F8 with a five hundred gallon tank and a three-stage pump. Myron Jones, in his Power Wagon pulled up in front of Lucy's, "Come on boys let's go!"

Johnny, Paul, and Roy jumped in the back and were gone.

Lucy came out of the crowd and organized the Women's Club. "They'll need coffee and sandwiches before this is over. Come on back to the kitchen and we'll get it going."

As the pumper roared up the hill, the plant came into view. It had three distinct parts, the ore house and crusher to the east, the production area to the west, and a fifty foot, diagonal galley connecting the two. The ore house was an inferno. Fire poured out of every window and flared out of holes in the iron siding. The corrugated roof glowed red and had started to warp. The snow bank around the ore house had disappeared.

The pumper lurched to a stop below the plant.

"Oh my God," whispered Brad Fagan, the volunteer fire chief. He held his watch up to the dashboard. It was eight-thirty. A line of pickups loaded with volunteers pulled up behind the pumper. All hands gathered in the half light as embers rained down. Slim Webb and his crew ran from the control room.

"Is everybody out?" Brad yelled.

"Yeah, we're all here."

Brad turned to Myron, "You helped build this monster, what'd we got here?"

Never one to be hurried, Myron pulled down his mustache and pondered. "Well, I tell you, if she jumps to the production area, we'll lose everything. And, hum…there's a two thousand gallon propane tank over there. So, I'd say, you can kiss our ass goodbye if she blows. The lights are on, so we still got juice. That'll keep the electric pump going. And, what the hell was I thinking? Oh yeah, the water hydrants are over there below the feeder."

"Ok boys, listen up!" Brad shouted. "Myron's right, too late to save the ore house, got to stop the spread. Jack and Billy, hook up the squirrel tails from the hydrant to the engine pump. Paul, Johnny and Roy get on down to the river and light off the auxiliary pump. No telling when we'll lose power up here. Myron, you take Ronnie and Robert to the other side of the galley. See if you can get a hand line on that side, don't take any chances. We'll run a hand line from this side. Just try to contain it. Ok boys, grab a respirator and team leaders take a Motorola."

Brad got on his handheld and called the dispatcher, "Yeah it's big, need all the help we can get. Call the fire departments from Dolores, Cortez, and Telluride. Tell them we need help.

Johnny switched on his handheld, "Brad, this is Johnny, the intake on the pump is all iced and the pump is buried in snow. It'll be a while."

"Okay, keep at it. Anything else?" asked Brad.

"Yeah, keep an eye on the galley. If she runs up the galley, she'll spread to the other side."

"Got any ideas?"

"See if you can get a hand line to the top of the galley. You'll have to come up the catwalk on the reactor. Throw water down the galley. It's framed in sawmill-grade lumber and probably treated with linseed oil. It'll damn well burn."

"Thanks, let me know when you get the pump going. We got hoses running right up against the high voltage transformers. No telling when we'll get fried."

Myron and his guys had worked their way to the backside of the ore house and had a full stream of water pouring in a window. Outside, the sheeting buckled. Inside, the fire raced up the wood framing and onto the rafters.

"Brad, this is Myron, yeah we got water, but it's hard to get to the base of the flames."

"Same here. Can you get inside?"

"We're trying to pull off the siding, but it's damn hot," Myron replied.

"Keep the hose on it."

By ten that night, the Women's Club had set up a fire camp below the plant. Pots of coffee perked on a camp stove, surrounded by mounds of sandwiches. Dozens of hand lines now ran from several pumpers. Firemen came and went into the darkness. Exhausted crews in soot-stained gear stumbled back into camp, relieved by out-of-town volunteers in spotless yellow jackets. The smoke and fumes came and went with the shifting wind. Everyone wanted to help. Everyone was on edge. Everyone was in danger. Was dynamite stored inside? Which way would the ore house collapse? Were the chemical fumes toxic?

Chester arrived later and went right to work pulling the hand lines away from the electrical transformer. Water sprayed from every hose coupling and washed down around the base of the transformer. He heard Stanley Pritchard screaming before he saw him. "How did this happen?"

Stanley held a double-bladed axe by the top of the blades and jammed the handle into the mud.

Chester stared in disbelief.

"How did this happen?" Stanley screamed.

"Let's get the fire out, for now."

In a frenzy, Stanley pounded the axe handle harder and harder. Blood poured down the axe handle, "I want to know who is responsible?"

Chester calmly wrapped one arm around Stanley's shoulder and took the axe away. He held him motionless for several seconds debating what to do. This is not the best time, he considered, but he needs to hear it, "What is wrong with you? You've got volunteers from Cortez to Telluride risking their lives to save the plant and you don't give a shit about them, just looking for someone to blame."

Stanley stared blankly at the blaze.

Chester finally released him, "Go on up to the fire camp and sit. We'll talk later."

The ore house was now enveloped in flames. There was no saving it. With a handheld to his ear, Brad Fagan directed the firefight from the hood of the Big Job Pumper.

"Johnny, this is Brad. You got that pump going?"

The gas-powered auxiliary pump sat on a concrete slab just above the river covered in snow. The six inch diameter intake hose ran down to the river. Johnny's hands were frozen to the hand-pull starter rope. Paul fiddled with the choke. Roy was knee deep in the river trying to clear the intake. They were covered in ice and shook violently.

Johnny spoke in short bursts, "Hell no…got gas…got spark…just won't turn over."

"Johnny! You've got to listen to me. I've got a crew coming down to relieve you. Get your ass back here, now. All of you."

Brad shook his head in frustration and switched channels, "Hello Myron, we've got a crew coming in to relieve you. How's it going?"

"We pulled off some siding. They got oil drums stacked in the corner."

"Ok, listen, do not go inside. I repeat, do not go inside. See if you can get some water on the drums, keep your distance."

As Brad conferred with the other fire chiefs alongside the Rico pumper, the wind shifted to the south. They disappeared in the smoke. Brad hit all channels on his handheld, "This is a message to all crews. We just got a wind shift to the south, I want you all to pull back, I repeat pull back, roger my last."

Minutes later, the roof of the ore house caved in. A smoldering heap of iron sidings, crusher parts, and twisted conveyors remained. The smoke cleared for a moment and the summit of Telescope Mountain appeared on the skyline. An indifferent moon loomed above it.

At two-thirty in the morning, Brad called it quits. The ore house was lost, but the rest was saved. The volunteers straggled back to the fire camp, feeling a small sense of triumph. No one asked how it started, yet everyone wondered. Lucy announced she was serving a free breakfast down at her place. Brad stayed behind with a skeleton crew. He needed to write a fire report.

Maggie brought him a cup of coffee, "Thanks darling, see if you can find Slim, he was up here when it started."

Paul, Johnny, and Roy huddled around the fire camp stove, wrapped in army blankets. The color had returned to their faces. Johnny rubbed at a lump of pyrite he'd picked up in the yard.

Brad walked over, "You boys did your best. I don't think the auxiliary pump had been lit off for years. Go get you some breakfast. And Johnny, I got a couple of questions for you."

"Yeah, maybe later," he said without looking up.

Paul waved Brad to the side and lowered his voice, "Say Brad, do you mind if we take a look at what's left of the ore house?"

"Go ahead."

As they stepped inside the shell of the ore house, heat radiated off the tangled metal. Paul and Roy sat on a concrete piling, enjoying the heat. Johnny picked his way to what was left of the ore crusher and then struck the iron jaws with the pyrite lump. Sparks flew. Paul and Roy watched without a word. After a long silence, Roy pulled out what was left of a half pint he carried with him when he performed. He had a sip and passed it to Paul, "I tell you what I think, I think we ought to finish her off right now. One stick of dynamite under the propane tank, that'd do her. I got me a mo-bile home, you know. So, I am mobile."

Johnny stared blankly at the jaws of the crusher.

Paul got up to go, "Come on Roy, let's get us some breakfast."

"What's with Johnny?"

"I think we've lost him for now. He's out there somewhere, trying to see around the corner. He needs to be alone."

As they got to the pumper, they looked back. Johnny was surrounded by flying embers. The wind had kicked up.

Chapter 16

BURNT OFFERING

Covered in a mantle of ash, Johnny slumbered fitfully in the crater of the ore house, wrestling his demons. He awoke with a jolt as an ember landed on his hand. Crushed pyrite smoldered in all directions. A blood-red sun shone dimly through the smoke and fumes.

Maybe he'd passed to the other side. He had hoped for better. Wiping the soot from his eyes, he checked his watch. It was six in the morning. How about that? He'd lost three hours, and the bad dreams had returned.

This time he was stuck on a rock in the middle of a flooding river. Either way he jumped, back to where he came from or forward to new ground, there was risk. And even if he got the direction right, he still had to get the jumping right. If not, he was a goner, swept under a rock, pinned on the bottom. Like all his dreams, he was alone, in danger, and forced to make a choice. And there was nothing safe about staying put, not on a rock in the middle of a rising river. For years he'd ignored the dreams, witches in the graveyard, spooky but meaningless. Just shook them off, stuffed them in a gunny sack, and pushed them in a dark corner of his head.

Lately, it wasn't working. The dreams kept coming and they were getting scarier. Anymore he wasn't sure of what was real and what wasn't. There was the whole business of staying or leaving West

Virginia. And there was the staying or leaving Rico, making it as a foreman, or cracking some heads. And now there was this fire, and his report, and Brad pushing him to step forward, and to think that Dorothy had predicted it with her homemade tarot cards. And Polly, fed up with moving. And there was always Roy to deal with. Good or bad, right or wrong, stay or go, what a damn mess.

They'd all gone to breakfast hours ago. He'd better hurry. As he got to moving, the schemer in him emerged. He sensed he had an edge, but wasn't sure how to play it. It would cost the Argentine at least two hundred thousand to rebuild and refit the crusher and ore house. Five percent of that would have solved all the problems. It just didn't add up. Hell, maybe it was one of those phony insurance deals. He could have pushed Chester harder. He probably should have.

A blanket of smoke hung over the town. As he turned up Glasgow Avenue, sedans and pickups lined both sides of the street. Three fire trucks were parked in the lot next to Lucy's, and several horses were tied to the fence in the back. That's a lot of pancakes and beer he thought. As he got closer, he heard a painful screeching, like alley cats going at it. It was Roy and Tanya doing Elvis. Nothing was very funny, but that was funny. They were awful.

Volunteer firemen from three counties were packed inside. Red eyed and weary, they gulped down breakfast and tried to sort out the night. The crew from Cortez, at one table, dug into their second stack and debated how it had got started. At another table, cowboys up from Stoner, sipped their whiskey and hoped the fire would shut down the whole damn place. By now, none of them were feeling any pain. Luther, one of Leon's sidekicks, asked him for the tenth time, "So, how hot was it up there last night?"

Leon pushed back his Stetson and grinned, "Hotter than a whore house on nickel night." They broke up laughing. Further down the bar, Argentine workers wondered if they'd have a job come Monday. To everyone's relief, Roy and Tanya finished their duet. Myron Jones, the Mayor, jumped up on the bar, in his standard gear, plaid shirt, rubber boots and Levis, all filthy black. A cockeyed grin flashed below his mustache, "It was one hell of a night, God bless you all," he lifted his glass in a toast. A cheer rang through the bar, as coffee cups, beer bottles and shot glasses were raised. "Here's to all you volunteers, you did a great job." He nodded to each table. "To our friends from Dolores, to our friends from Cortez, to our friends and neighbors right here in Rico, and a special thanks to our fire chief, Brad Fagan, and, to our friends from Telluride, the second most important mining town in the San Juans."

Leg weary, Myron sat on the bar. "And finally, let's toast the Rico Women's Club who kept us going through the night, and for this wonderful breakfast. Somebody get 'um out here!"

Lucy, Maggie, Polly and all the rest marched out from the kitchen in their aprons, and took a bow, followed by another round of cheers. It was a fitting end to a miserable night.

When things died down, Myron held his hand in the air for silence, "Drive carefully and let's all thank God no one was injured."

After a moment of silence, they all headed home. Through it all, Johnny stood in the back out of sight. As folks were leaving, he walked over to Polly, "Hey girl, nice job."

"There you are. Everyone's been asking for you."

They embraced, "I know, come on."

Brad Fagan caught him going out the door, "Got a minute Johnny? I need to finish my report."

"Later."

"Ok, but I got to make my report to the town board Monday night."

Johnny's eyes narrowed, "Fine, but not now."

"We can talk tomorrow. You do need to know, that Slim said it started from a cigarette, somebody from your crew."

*"I am a roof-boltin' daddy, I am a roof boltin' man,
Light on my head, stoker in my hand.
Putting up pins just as hard as I can,
I am a roof-boltin' daddy, I am a roof boltin' man."
—Gene Carpenter*

Chapter 17

SILVER CREEK

They were blowing hard when they reached the second switchback. The town spread out below them. The Fagans were on a Sunday hike up Silver Creek with the Carnifaxs.

"Come here, Storm," Polly yelled as she eased herself down on a bolder. The morning sickness had started. Brad Fagan spread an old claims map on the ground and pulled out his binoculars.

"See the corral just south of Silver Creek?"

Johnny took the binoculars, "Okay, sure. What's that next to it?"

"It's a stone slab, all that's left of the old powder magazine. They say in the forties a miner, down on his luck I suppose, lugged a keg of powder from the magazine out beyond the corral, sat down on it, lit his pipe, then lit the fuze, and boom."

Maggie caught Polly's eyes, they both shook their heads, "Can we move on?"

"Who owns it now?" asked Johnny.

"All that area down there and just about all the claims above here, all the way to the summit of Telescope Mountain are owned by Rico Argentine."

Johnny looked to the north side of Silver Creek, "How many are working mines?"

"Not many, we'll see lots of abandoned portals further up, and come spring you'll see acid drainage running out of them and down to the creek. It will go on forever."

As they continued up the trail, Johnny wondered if it were this bad everywhere, or if Brad was slanting things. He had a lot of coon dog in him and couldn't be put off completing his fire report. He sure disliked the big mining operations. Storm trotted up the trail then curled up on the side of a clearing. Johnny followed.

"Hey Brad, why is it so barren over there on the lower slopes of Telegraph Mountain?"

"Come on Johnny, connect the dots."

Maggie smiled, "Okay professor, get on with it."

Brad carefully took off his glass and wiped off the sweat, "The fumes from the acid plant killed the trees, then folks in town cut them down for firewood."

"And," Maggie added," we've all had it burn our lungs and rust our roofs. And guess what? It also kills off the trees."

Polly elbowed Johnny in the ribs, "Smarten up boy."

Johnny threw on his pack and headed up the trail, "Come on folks, it's a wonderful day. Let's enjoy it."

Storm bolted up the trail, as the three exchanged furtive smiles. They had an agenda and were far from finished. At a sandstone outcrop, Johnny studied the view across Silver Creek. Brad was back down the trail, panting heavily.

From the outcrop, Blackhawk Mountain cut a hard edge across the skyline. To the right, Dolores Mountain ran down to Newman Hill. Its stark granite summit gave way to fields of green, variegated stands of spruce and pine, which ended their run in aspen groves. Passing clouds splintered the midday sun, creating subtle shifts of light and shade. Only the abandoned mines spoiled the alpine scene. Below gapping portals, towering gray mine dumps spilled down the mountain. Rusting mills and rail spurs lay cluttered nearby. Even the most unrepentant mining company would feel some remorse at this scene.

Brad and the ladies stopped at the outcrop.

Polly turned to Maggie, "You'd think they clean up their own mess, wouldn't you?"

"Mother Nature has a way of reclaiming what man has messed up, but this may be beyond the old girl."

Johnny looked out wistfully, "They must have been some tough old boys, down in the mine all day, sleeping in a bunk house hanging off a cliff, tough life."

"Yeah, some of them are still around. They come up for the Fourth of July picnic, mostly to meet old friends and tell their stories. Okay gang, let's go. We've got another hour before lunch," Brad announced.

Maggie pulled a blanket from her pack and spread it on the ground, "Sit down Polly we're taking a break."

Puffing heavily Brad collapsed. "That is one hell of a good idea, darling."

The air had gotten thinner as they made their way up the trail. Brad's altimeter read nine thousand four hundred feet. For now, they were content to take a break and enjoy the moment.

"So Johnny, tell us about old John D. Carnifax in the Civil War."

"John D. was an ornery cuss, hated the Yankees. He refused to carry Union soldiers across the Gauley River on his ferry. He told them the water was too high."

"What'd the Yankees have to say about that?"

"They loaded the ferry with soldiers and horses and off they went."

"So what did John D. do?"

"When they got to the middle of the Gauley, he cut the cable. They never were heard of again."

"Damn, I like that one. So, would you say the Carnifaxs have always had an ornery streak?"

"You might say that."

Polly nodded in agreement.

Johnny stretched out on his back and watched the clouds float by, "Give young Brady our best. He's had quite a season."

Maggie passed elk jerky around, "Yeah, they made it to the regionals this year, got all their players back. Next year, they're going to state."

Brad turned to Johnny, "We hear from Polly that you were quite the football player in high school."

"Nothing special."

"Said you were state champs."

"That's my boy," cooed Polly.

Johnny lit up, "Now that was a dirty game, nothing worse than a gorilla and a home town ref. I got benched. This line backer was off sides all night. I'd go back to pass and he'd hit me, even when I got the pass off he'd hit me."

"What'd ya do?"

"I finally slugged him. Ref calls it. Coach benches me, fourth quarter we're behind by four points, less than a minute to play. We got the ball at midfield."

"And you're ridding the pine?"

"So, I walked up to coach. Everybody is yelling, "Johnny! Johnny!" in the stands. He gives me a dirty look and says, "Get in there, throw the bomb."

"Long bomb yeah! What else," Brad started clapping, "Yeah, go Johnny!"

"I went in, called the play, got up to the line, and gorilla boy starts screaming, "Pass! Pass! Everyone in West Virginia knew what we were up to, and...," Johnny paused, flashed his teeth, and started chuckling.

"Come on, what happened?"

"Called an audible, ran right over gorilla boy's fat ass, fifty yards for a touchdown. That did it, state champs!"

Brad jumped up and threw on his pack, "Johnny boy, we got a big game tomorrow night at the Town Hall and we'll have a few gorillas swinging from the rafters."

Weary of the prodding, Johnny moved on up the trail. They hadn't gone more than ten minutes before they came to a long, clear-cut bench running along Silver Creek.

"And what do we have here?" Johnny asked.

"That's the Blaine Mine. The Argentine started mining lead and zinc there in 1937. There are miles of tunnels running way under Blackhawk Mountain. That's the flotation mill, processes one hundred and thirty-five tons of ore a day. They got real busy during the Korean War when the prices of zinc and lead shot up. Down below on the track they are dumping tailings into the Silver Creek."

Mine workers dumping tailings into Silver Creek.

"That can't be good," Polly observed.

"It's not. It flows into the Dolores River and adds to the contamination," said Maggie.

"And what is that nasty sea of sludge below it?" Polly asked.

"That's the settling pond, where nothing stays settled very long. It leaches and on occasion overflows into the creek."

Johnny could feel the burn move up his neck and across his face, "Okay gang, hang on. Have any of you been two thousand feet down in a mine? Have any of you rode a jackhammer for a ten-hour shift? Hell, it used to take us an hour to get from the mine portal to the coal face. You work in the dirt, you're gonna get dirty."

Polly moved to his side, "Easy Johnny."

"Have any of you ever seen a mine you approved of? I guess you're waiting for an angel band in long white robes to march out of the mine and sail on up to heaven in a gilded ore car."

It got prickly, like barbed wire strung between them. No one spoke for several minutes. Finally, Brad broke the silence.

"Look, I don't understand mining like you do, but I've been around it. My dad was a miner up here in the twenties and thirties. He died an early age, tuberculosis or something."

Maggie tossed out her blanket, "This could be one short hike. Let's settled down and talk this thing out."

She leaned her full bodied frame up against a tree, pulled off her bandana, and shook her hair free, "Brad's right, none of us know mining like you do, and it is a dirty business, and that's what we're trying to clean up."

She paused, gathered a head of steam and continued, "Just look at that hillside, pockmarked with mines and tailing. That's not right, and what's happening down at the acid plant is sure as hell not right."

Polly nudged Johnny, "She something isn't she?"

"Yeah, she's something. She talks like she grew up in a coal camp."

Brad pulled a notebook from his pack, "I've got to finish the fire report for the meeting tomorrow night, and I need to figure out how it got started. I suspect it was negligence on the part of the Argentine."

"Could have been," Johnny shrugged.

"Listen to him. Will you?" Polly urged.

Brad flipped through his notes, "I interviewed Slim, he said, someone on your crew was smoking in the ore house, says that's how the fire got started."

"Is that a fact? So Brad, tell me, have you ever been in the ore house when it's running? It's the loudest dirtiest place in the plant. If you take a smoke break, you go outside."

"Come on Johnny, we've heard you did a report on the fire hazards in the ore house. That would be very useful tomorrow night."

"Where did you hear that?" Johnny eyed Polly and Maggie. "Your Women's Club, is full of snitches."

Polly moved between them, "Johnny, don't you understand, this is our chance to change things around here."

"I know you don't believe it, but I am on your side. But please, do not underestimate these guys. Sure we may have a slim edge here, but if we pull out my report, they'll have some flannel-mouthed lawyer twisting it around in seconds. And Polly and I will be heading back to West Virginia. Let's go slow."

Polly flushed, "Like give up, like the old boy that ended it on the powder keg?"

"Look, they probably figured we'd make a copy of the report. So, before we start flashing it around, let's hear what they got to say. Sometimes, change can come from helping the bad guys when they're down."

"Johnny, do you know how sick that sounds?" Polly shouted.

Rico Town Hall

Chapter 18

TOWN HALL MEETING

Winter 1963

The Rico Town Hall is an imposing brick and sandstone structure. Back in the 1890's when silver was booming; it was the county seat, serving as the Dolores County Courthouse. But as any local will tell you, the bean farmers from Dove Creek stole it away in 1946. Though still a great source of civic pride, it is now the town hall. The City Clerk's office is in front, on the ground floor where everyone came to pay their water bill. The library is off down the hall. Serious criminals are locked up in Dove Creek; harmless drunks sleep it off in the basement jail. The centerpiece of the town hall is the courtroom on the second floor. Long arched windows run from the floor to the high vaulted ceiling, offering the best view in town. The courtroom has seen its share of social events: card parties, cake sales, pot lucks, reunions, receptions, and dances for every season. On the civic side, the Rico Board meets once a month in the courtroom.

By 7:25 that evening, the mayor and the rest of the board had settled behind an elevated oak bench that extended across the front of the court room, set off by three arched windows that looked out on the darkening skyline. The American flag stood to the left, the Colorado State flag to the right. Rows of wooden benches extend to the back wall, divided by a central aisle that served to separate the political intentions of those that gathered that evening.

Board meetings in Rico were generally a sleepy affair with few in attendance. Not tonight. The fire at the plant was on the agenda. The right side of the aisle was packed. Chester Ratliff and two company lawyers sat in the front row and fifty or more Argentine employees and loyalists spread out behind them.

A smaller group of concerned citizens sat on the left. The Fagans, the Spitzers, and the Carnifaxs were in the front row. Maggie sneezed repeatedly into a Kleenex, while Storm slept on the floor next to Polly. Hartley and Irene Lee, Lucy Fahrion, Jim Rychtarik, sat in the next row. Hartley, a local merchant, was also a well intentioned muckraker for the *Dolores Star*. Behind them sat Roy, Tanya, Chee, and Trish Vanderville. Leon and several other cowboys from Stoner were spread out behind them. The volunteer fire chiefs from Dolores, Cortez and Telluride sat further back.

Mayor Myron Jones rapped his gavel at 7:30, "Let's get started. We've got a full agenda and a full house, so keep your comments short."

Myron was an independent miner, with his own claims, his own mines and worked them when it pleased him. He didn't expect much to come out of the meeting. The Argentine had their lawyers and pretty much ran the town. There had been several fires at the plant lately, but their fire insurance always seemed to cover the damages. The ore house fire could have been prevented with a few simple precautions. That was common knowledge. Maybe Brad Fagan's investigation would uncover that.

The minutes of the last meeting and the Treasure's report were read and approved. Myron pulled down on his mustache, scanned deliberately across the room, "All right, let's get on to new business. Mr. Chester Ratliff from the Rico Argentine has asked to make an opening statement."

Chester stood up and smiled broadly, "First off, I want to express my appreciation to the Town of Rico and to all the volunteer fire departments for their help in controlling the fire at the plant, well done."

Hartley Lee jumped up, "That's the third major fire in three years. What are you doing to prevent them?"

Myron tapped his gavel gently, "Calm down Hartley, you'll get your turn soon enough."

Chester conferred with his lawyers and then went on, "I'd like to ask each of the fire departments, to make a survey of the damaged equipment from the fire and submit it to me. Rico Argentine will pay to replace all of it."

As if on cue, the right side of the aisle erupted in applause. The left side remained cautiously quiet. The room was as divided as the Dolores High School gym when the Bears played the rival Telluride Miners in basketball.

Jim Rychtarik, who couldn't hear his phone ring anymore, stood and waited to be recognized.

"Ok Jim, keep it short."

"I've got no report."

"I said, keep it short,"

"I intend to, quit your gabbing. Myron, you know I am a regular at these meetings, and I'd like to know where Mr. Stanley T. Pritchard is tonight. He's the one who stirs up most of the problems around here and then he sends Chester down here to sweet talk us."

"Thanks Jim, it's been so noted. Go ahead Chester."

"In conclusion, I am pleased to say, there will be no lays offs or lost time. We've got three to four weeks of crushed pyrite in the reserve ore bins, and we've resumed operations. Until the ore house is rebuilt, we'll load it manually. We'll all work together on this. It's a partnership."

Heavy applause from the left side of the room followed.

Chee turned to Roy, "Partnership?"

"It's like a team."

"Is partnership, another *bilagaana* (white man) word, like treaty?"

Johnny felt Chester's words were encouraging. Maybe the Argentine was ready to make some changes. He whispered to Polly, "Sounds promising."

"Sounds like a whitewash to me, Chee will understand."

Myron feathered his gavel, "All right folks, let's quiet down. Hartley you're next, then Brad Fagan."

Hartley stepped to the front of the room and faced the board. "Let's have some straight talk. The fire is just one more example of a poorly run operation, nothing changes. Acid fumes are choking the town and the fish are dying in the river."

Suddenly the floor began to shake. Leon and the Stoner cowboys were stomping up and down in the back of the room, accompanied by the jingling of spurs. Then they started to chant, "The water stinks, cattle won't drink, the water stinks, the cattle won't drink!"

Johnny looked back at the cowboys and then over at Roy, who wore an evil grin. He was popping his knuckles, his normal pre-fight warm up.

Myron banged away on his gavel, "Leon, hello Leon, this ain't no rodeo. Quiet down or I'll have Robert here throw you in jail."

Myron nodded to Robert Tate, the Town Marshall. Robert never missed a meal, was known around town as Full-Plate-Tate. He was armed with a pistol, a night stick and a nasty disposition.

Chee smiled at Leon.

Johnny nudged Polly, "It's not often you see cowboys and Indians getting along."

Hartley, who was itching for a showdown, continued, "My question for the board is, can the Town of Rico do anything to correct these problems?"

There was a long pause, and then Mr. William Thompson, the Town Attorney, stood up. He was all starched and pressed and tidy, could have been a preacher or an undertaker. A shock of white hair fell over his collar.

"The town of Rico has the right to get an injunction against Rico Argentine Mining Company, which would require court action. Any citizen who has been affected by the existing conditions has a right to get an injunction through court action. Then the town of Rico could get a written commitment from the Rico Argentine that they will not let happen, what has occurred in the past forty-eighty hours, namely fires and pollution."

The right side of the room gasped. Chester quickly huddled with his lawyers. Then one of the lawyers, a Mr. Richard Parma, stood up. He faced the board and flung his arms open, "Honorable members of the board," then spun on his heal and faced the audience, "and worthy citizens of Rico, as Mr. Thompson is fully aware, Mr. Stanley Pritchard, the President of Rico Argentine, has responded to these issues in writing and I quote, "every available precaution has been taken with the plant to guard against fumes and other disagreeable results and to provide for the safety of the workers and those in the immediate vicinity. Unfortunately, there have been, on occasion, human errors. Some inconvenience might have resulted to some of the inhabitants"."

Polly jumped to her feet, "How dare you? "Every available precaution has been taken?" and… "unfortunately on occasion, human errors…"

She paused, gathered steam, and turned to Argentine supporters, "Do any of you believe this? I don't think so."

Even the most loyal supporters remained silent.

Johnny tugged at Polly's arm, "I thought you were going to cool it,"

Myron stood up and waited until it was quiet, then proceeded calmly, "Let's have no more outbursts. Polly, please sit down."

Smiling smugly, Lawyer Parma proceeded, "Let me be clear, madam, there is no statute in Colorado, governing the contamination of air."

Chester again conferred with his lawyers and then rose to speak, "Look, no one is denying that there have been problems and for that we apologize. We're part of the defense industry and we are in the middle of the Cold War. But make no mistake, precautions have been taken. We want to get this right, for our sake as much as for yours. My only request is that when you have a problem, you come and talk to us."

He paused and glared at Hartley Lee, "Don't dash off a misinformed article in the *Dolores Star*. Now on the subject of precaution, let me say this. The Public Health Service inspects the plant three times a year and submits a report of their findings. They'd shut us down in a heartbeat, if we weren't up to snuff."

The left side of the room was incredulous. Maggie Fagan jumped up, "Myron, can I say something?"

"Myron raised the gavel into the air and exhaled, "Maggie, you're out of order, Mr. Thompson's up, please sit down."

"Your honor or whatever, you are out of order. I am not sitting down. I got something to say."

Sheriff Tate lurched up the aisle and whispered, "You need to sit down, dear."

Maggie stood her ground. Sensing the tension, Storm jumped up and started growling at the sheriff.

"All right Maggie, say what you've got to say, and somebody put a leash on Storm."

Never one to lose a dramatic opportunity, Maggie marched forward. Pausing in front of Chester, she handed him a fresh Kleenex from her purse, "Boo hoo," she whispered. "Has this been hard on you?"

Everyone got a kick out of that. She came to a full stop in front of the American flag, cast a withering eye at the board, and smiled inwardly, it's time I made a stand, we can get by on one salary for a while. She turned to the gathering.

"Good evening, let me ask, has anyone in this room seen the Public Health Service inspectors in town?"

Both sides of the room looked around blankly.

"Come on folk don't be timid. You couldn't miss them. They travel in a convoy, six doctors, three trailers to live in, another as a doctor's office and a lab, and a mobile X-Ray truck. How about it Lucy, have you seen any of that in Rico? They say you know about stuff before it happens?"

Lucy smiled and shook her head. Lawyer Parma jumped up and started waving his hands in the air. Myron gestured with his open palms to sit down. The room was buzzing with tension.

"All right then, let me finish," she glared at Chester. "Now, you all know that I work in the Burley Building for the Argentine. Right, and I know a thing or two about what goes on up there," she paused and the room froze. "I handle all the filing and paper work, and I can tell you straight, there are no inspection reports."

"The girls got grit, but she may have got ahead of herself," Johnny whispered to Polly.

"Yeah, what happened to your grit?"

From the back, Trish Vanderville started chanting, "Power to the people!"

The Women's Club was making quite a show of it.

Lawyer Parma bolted to his feet, "Your honor, your honor, this is hearsay, it's inadmissible."

"Lighten your load there Mr. Parma," Myron instructed, "this is no courtroom and I am just a humble mayor. If you've got inspection reports, bring them forward."

Lawyer Thompson rose, "Earlier, Mr. Lee asked a simple question, "Can the Town of Rico do anything to correct these conditions?" And I believe we can. There is a state law governing the contamination of streams, and the enforcement, would require action. So yes, the town of Rico can do something. It can petition to state officials to take action."

Chester was up again, waving to be recognized.

"Go ahead," Myron sighed wearily.

"Let's be clear. The Argentine is not ignoring the problem. We have doubled the number of tailing ponds in the last six months and rebuilding the ore house will cost us plenty."

Myron pulled down slowly on his mustache. He was torn. He wanted to run a fair meeting, but he damn well knew the ore house fire was another example of Argentine's slipshod practices. He knew the Argentine had fire insurance, and wanted to get something in the minutes on that.

"Say Chester, you gonna get any help from the insurance company?" he asked casually.

Caught off guard, Chester replied, "We'd better."

The Argentine lawyers cut him off immediately.

Myron chuckled to himself, *got him.*

"Ok, Brad you're up, what can you tell us about the fire?"

Brad leafed through the pages of his notebook, "We can't say for sure how it got started. I've talked to a lot of folks. Everyone has an opinion, but nothing conclusive. I haven't finished my investigation yet."

Brad hoped Johnny would speak up on his fire prevention plan at this point.

"Ya think you're going to learn anything more?" Myron asked.

"Can't say, need to be thorough. Insurance companies will want a final report."

Chester nodded to Slim, the night shift supervisor at the plant. He'd been lawyered-up before the meeting. Slim stood up and shrieked in a dry reedy voice, "Tell 'um what I told you Brad, tell 'um!"

Brad flipped through his notebook, "Slim here, says it was started by someone on the day shift, smoking in the ore house."

Slim shouted, "I told you it was Carnifax's crew. He's the guy you need to be talking to."

Myron scratched his head in disbelief, "Hang on Slim, are you saying a cigarette started the fire?"

Slim recoiled then launched into his spiel, "Damn rights. There's pyrite dust caked everywhere, doesn't take much to get it going." Slim hesitated, caught in his own snare, "And you got oil drums and things like that around."

Chester flashed red as he tried to cut him off, "That'll do Slim, let me explain."

Like a slack tide, the room fell silent. This was a lot to take in.

Johnny's first impulse was to strangle Slim until the words fell into place. 'pyrite dust caked everywhere, oil drums.' He throttled back. They sure as hell don't want that in the fire report.

"Chester, you hang on. Brad, did the pyrite dust come up in the other interviews?" Myron asked.

"Most of them."

The floor groaned as Chee took his feet with a bandana around his forehead and fire in his eyes. He looked around the room until he found Slim, cowering at the end of a bench,

"Slim is wrong. The welders started the fire."

Slim jumped up. Someone yanked him down again.

Both Argentine lawyers were on their feet seeking to be recognized.

Myron tapped gently on the gavel, "Sit down boys. Brad, what do you think?"

"I checked. They had an arc welder in there around shift break, working on the crusher arm. That's probably what set it off. The fire hadn't crowned when we got there. We could still see pyrite dust caked on the cross-framing and the ceiling joists. That burned first, and got the wood going."

"Ok Brad, we need to wrap this up. Anything else?"

"The fire report asks if there were any previous assessments of fire hazards or plans for prevention," Brad's final opening for Johnny to come forward.

Johnny unfolded his copy of his fire plan and reread it. He was stuck, caught somewhere between fussing and scheming. Fussing, in this case, would be taking the fight to the enemy, going public with his plan. Before Polly and the baby, he would have gone right to fussing, yet, he was already on probation. This would get him fired. Scheming it would be, help the bad guys, become the good guys. The fires and breakdowns were preventable and made no business sense. The fire plan was his trump card, and Chester knew it. Better hold off for now.

"What about that Chester?" Myron asked.

"Look, we're an acid plant, not the fire department. Of course, we're always looking for ways to improve things. That's just good business."

"Anything specific?"

Chester knew he was stuck for the moment. If he acknowledged Johnny's plan, they would know he had not acted on it, and if it made it into the fire report, the insurance company wouldn't pay. Yet, if he said he had no plans in place, they would see that as irresponsible.

"What can I say, you've all got jobs. We're back in business."

Myron walked around to the first row of benches, "Hartley started this meeting by asking what the board could do to correct the acid plant problems, and then we heard Mr. Ratliff say we need to build a partnership and work together. I must say I see little coming out of this meeting that gives me much hope. Good night, folks."

Nothing much did come out of the meeting. The battle lines were drawn. The tension in town was withering, yet little changed. Lawyers on both side billed more hours and all manner of political finagling went on behind the scenes by the Argentine with the District Attorney and the Country Commissioners. The educated outlaws were on a tear. Perhaps a fissure of doubt had opened in the minds of the supporters, but not enough to prompt any action. The pollution disappeared for a while. Beyond a few rebels, most folks weren't interested in losing their job with the Argentine.

Chapter 19

Nihikéyah baah Dahaz' Áago, Bílá Ashdlá Ałdó baah Dahaz á.

"When the Land is Sick the People are Sick"

It was dark and pounding rain the morning Johnny got the call. Norman was having seizures and Chee needed help getting him down to emergency in Cortez. Johnny called Roy to cover for him, kissed Polly good-bye, and headed up the hill.

When he got to their *hooghan*, Norman was cradled in Chee's arms, his eyes were closed and there was a frightening rattle to his breathing. They moved quickly. Chee laid him in the back seat of Johnny's Ford Falcon and tucked a blanket around him. Johnny roared off down the mountain to Cortez.

Chee hadn't slept for days and was fighting the flood of feelings that washed through him. He pressed his temples and turned to Johnny, "The bad wind is burning inside of him. His time is near."

Johnny accelerated through the poorly banked Montelores Hill, not sure what to say, "We'll be there soon. They'll have something for him."

Norman's coughing broke the silence, as they sped down the mountain. He weighed less than a hundred pounds and his skin lay in folds on his face and neck. An hour later, Chee signed him in at the triage station in emergency. There were others before them.

Johnny and Chee paced the waiting room floor for over an hour until Doctor Chavez finally appeared. He wore a weary smile, "Hello, Chee. He's resting now."

The doctor laid a folder on the table where they sat. "I received his record from the reservation hospital two months ago. I thought I'd see you sooner."

"He wouldn't come," Chee shrugged and looked around. "This is Johnny, a friend of the family. So how is he doing?"

"Your father is very sick."

"He believes the bad wind is punishing him."

"Well, yes. He doesn't have much time," the doctor replied with sadness in his voice.

Johnny broke in, "Isn't there anything you can do pills, surgery, anything?"

Chavez opened Norman's record, "I am afraid it's a little late for that. We did X-rays and took a biopsy three months ago, but he never returned. The reservation hospital at Shiprock did an X-ray a year ago." He flipped back in the record, "And detected a mass on his lungs. They thought it was an old cyst or pneumonia. Here again, he never returned for treatment."

Chee nodded, "The Shiprock doctor only believed in white medicine. He went back to the reservation to find his *hozho*. For five days, the healer performed the Wind Way Ceremony, warding off the bad winds. He was at peace, for awhile."

A nurse burst in the room and handed Dr. Chavez an authorization. He scratched out his signature and turned back to Norman's records, "So, he stopped taking his medicine?"

"Come on doc," Johnny growled, "You sound like a lawyer. What's wrong with him?"

Chavez looked up wearily, "The technical name is oat cell carcinoma. It's a type of lung cancer. It usually comes from smoking."

Chee's face lit up and his hands started to tremble. He was on a knife edge struggling for control. Then it was gone. He calmly stated, "My father never smoked."

Johnny watched, feeling Chee's pain and his effort to restrain himself.

The doctor persisted, "Are you sure? This is critically important."

Johnny snapped. He slammed the table and shouted, "He never smoked. Are you trying to pick a fight?"

Two male orderlies in starched whites flew in the door. Johnny and Chee rose slowly and met their gaze. It was a face-off. The doctor waved off the orderlies, to their immense relief. Then he patiently gathered up the records that had spread across the floor.

"Gentlemen, I am sorry. But, there is a lot about the way this has been handled that bothers me. But first, let's deal with the patient. When I spoke with Norman today, he said he was having horrible headaches and great pain on the left side of his body."

Chee nodded his agreement, "He's always in pain, but keeps most of it to himself."

The doctor looked down at his notes, "These symptoms along with the X-rays indicate that the cancer has metastasized. We'll have to do more testing, but it appears to have spread to his brain. The sad, honest truth is, he's dying. You need to help him get his life in order. We can give you some strong sedatives that will help him sleep and something to relieve the pain. I am sorry."

Chee stared at the records, struggling to control that rush of feelings. The wail of an incoming ambulance echoed down the hall.

Johnny got up, "Can we see him?"

"He's in room three, down the hall. But let me warn you, he's not very lucid. He's out there on the edge, fighting his demons. He talks of the bad wind, and for some reason, of the Public Health Service (PHS)."

"What's that all about?" Chee asked, as they walked down the hall.

The doctor stopped outside Norman's room, "I am not sure. The last time he was in, I told him that the PHS was testing uranium miners. Somehow it stuck in his head. We need to talk before you go."

Chee nodded, and they quietly entered Norman's room. He was a faint shadow in the dark room. He sat on the bed hunched forward, knees crossed, eyes riveted on a ribbon of light that fell between the curtains. He chanted at the light, then spoke, "Now I see it, the two fires. The fire in the valley feeds my body and the fire on the mountain feeds my soul."

He collapsed into a fit of coughing, fought to clear his throat and continued, "Now, it is clear. In life, I must keep both fires burning. But, to die well, I must keep my soul alive. I must keep the fire on the mountain burning. Now, I know what I must do."

His eyes were glazed over, yet a smile shone on his face. He was at peace.

Chee put a hand on his shoulder, "Father, it's Chee."

Norman slowly drifted back from his vision, "Ah, my son, you are a great son." They embraced.

"And who is with you?" Norman rubbed his eyes.

"It's Johnny. He drove us down here."

"The crazy one? Bring him closer."

"Johnny, you are good, but you are a crazy driver."

Johnny shrugged, "We were in a hurry."

Norman beckoned them closer with open arms, pulling them into his spirit world, "Come, we must speak in confidence. Soon, my body will go cold, but I must keep the fire in my soul."

Chee clenched his teeth. A childhood images flashed through his head. A boy and his father at sunset, racing their horses up the mesa, the first snowfall on the high peaks, searching for lost sheep, planting corn, laughing at the moon, laughing. Chee steadied himself against the bedside.

"Look deeply into my eyes. Can you see my soul? This is no time for sadness. You must help me die well. Nihikéyah ba̧a̧h Dahaz Ą́ago, Bílá Ashdalá Ałdó ba̧a̧h Dahaz ą́. I must fight the bad wind that burns through me." Then in a faltering whisper, "Take my record to the Public Health Service"

"But why?" Chee asked.

Norman struggled to clear his throat, "To do what is right, to cure the land."

Johnny was also puzzled. Then, in a flash, it was clear. I'll be damned. Norman has solved the riddle of the mountain. He has arranged the turquoise stones into one, bold-ass-plan. He will not go quietly, but as a warrior. Johnny shook with excitement. His mind rushed ahead.

Chee nudged Johnny, "He wants to give his records to the PHS."

"He wants to do what is right," Johnny said softly, then turned to go, "Give me a second. I've got a great idea."

Johnny found a phone booth down the hall and called Maggie. She had mentioned the previous day that her brother, Michael Burns, was working in the area. Michael was with the PHS. Strange coincidence, Johnny thought, or maybe this was a sign as well. He was starting to think like a Navajo.

"Hello Maggie, Johnny here. Maggie, we got ourselves a situation here with Norman, and I am gonna need your help." Johnny quickly filled her in.

"Here's the deal," Maggie shot back," Michael's in Uravan today, but he's leaving for Grand Junction after work. I'll try to catch him before he leaves."

Johnny carefully avoided any mention of how he planned to get the medical records.

"Thanks Maggie, you're a dear, and Maggie, please call Polly. Tell her I love her, but I won't be home for dinner."

Maggie cackled, "Nothing new there. And Johnny, you be careful. You've got some driving to do."

Johnny smiled and hung up. It was over ninety miles of twisted mountain road to Uravan. He jingled the car keys in his hand, as he rejoined Chee in hall outside Norman's room. This was his kind of driving, "I got the PHS part figured out. Let's see about the records."

Chee stopped abruptly, "Slow down Johnny, we need to talk."

"No, we gotta go, for this to work."

They stood face to face in the room, frozen. Chee then guided Johnny to an empty corner of the waiting room.

"My people, the *Dine*, have had many agreements with the government, with the Navajo Agency, and all these damn bureaus. And most of them have been broken. So, why should the bureaus help us?"

"Damned if I know. The doctor says they are doing studies, and Norman had this vision. But, it's your call, chief."

Chee towered over Johnny. "There is danger here. The roads are bad. It's storming in the mountains. And I guess, you plan to steal the records?"

"Borrow, just borrow."

"So, there is danger in this?"

Something snapped in Johnny. A fire flashed in his eyes and a snake was in his throat, "Hell yes!" he hissed. "Come on Chee, this is just a poor, white boy talking to you. I haven't had much luck in my life, not with unions, not with the bureaus, and not with any slippery-ass, government agency. So, don't think I am some sort of good-will ambassador."

"No Johnny, you are not a good-will ambassador. You are Crazy Bear," Chee smiled, "and you are on fire."

They both smiled.

"And you, Chee, are Big Thunder, so let's hear you roar."

Johnny paused and looked deeply into Chee's eyes. "Look, all I know is, if my father made a final request of me, I'd walk on razor blades to get it done."

Chee smiled slowly, "So, what are we waiting for *bilagaana* boy?"

They headed for the doctor's office, who was off on another emergency. Chee stood guard in the hall, as Johnny searched the office. He found the medical records in a filing cabinet behind the desk. The first drawer went from A-F. There it was under B, Benally, Norman Benally. As the doctor rounded the corner, Johnny rejoined Chee.

He pulled off his rubber gloves and invited them in, "Thanks for waiting. Now, this is in strict confidence, and I don't totally understand it myself, but here goes. U.S. Vanadium in Uravan, where Norman worked for years, has subpoenaed his medical records. It's very strange."

"But why?" asked Chee.

"I'm not sure. But it's happened before. It's like they want to know what we have diagnosed, what's in the records. And here lately, and this is strange, the Public Health Service has also made formal requests for some of the same records. Something's up."

"So, what will happen to Norman's records?" Johnny asked.

"Yesterday, our lawyer told me that we'd be in contempt of court, if we didn't hand them over to U.S. Vanadium. We'll see. Until then, they stay right here."

The wail of an ambulance cut through the ward. The doctor's phone flashed red. "Yes, I'm on the way. Sorry gentlemen, I've got to go. You're free to take Norman home, just check him out at the front desk. Good luck."

They loaded Norman in a wheel chair and pushed him to the front desk. The sedative had worked, and Norman dozed peacefully.

"I'll get the car and meet you outside," Johnny said and disappeared.

Chee looked at Johnny curiously, "Where's your coat?"

"Ah damn, I forgot it."

He returned to the front desk and nodded at the administrator, "Do you mind, I left my coat in Doctor Chavez's office?"

Just then the phone rang. She shook her head in frustration, then smiled at Johnny, "Go ahead."

Johnny was panting as he pulled the Falcon into the entrance. They loaded Norman in the backseat and raced off into the pounding rain.

Chapter 20

NINETY MILES OF BAD ROAD

Johnny skidded off the blacktop onto the county road. He smiled broadly as he fishtailed in the loose gravel. The rain pounded the roof of the car.

"You know where you're going?"

"Sort of," Johnny gunned it up a steep section. "I had a word with the ambulance driver. The best road to Uravan is around through Dove Creek and Slickrock, but it's closed with a mudslide. We're on the back road, runs up to Groundhog Reservoir and down through Disappointment Valley."

"Disappointment Valley?" Chee winced. "You think it's open?"

"Maybe. There's some small creeks and dry gulches that can get flooded. We've gotta fly."

Chee looked back down the road, "Nobody's after us yet, but it won't be long. Once Chavez gets back in his office, he'll figure it out."

Norman bounced in the backseat like an old cowboy on a fresh horse, as the Falcon bucked over bumps and potholes. His face was lined with pain, but he hadn't lost his grit, "You trying to break me, Johnny?"

Johnny grinned in the mirror, "Hang on Norman, we'll get you to the Uravan."

"Tie me in the saddle, if you have to."

Johnny accelerated through each switchback as they climbed the grade. They raced across the sage and juniper foothills and up into the big trees, sliding from side to side on each turn. Then they were on a straight stretch, wash boarding through ruts, bouncing off the oil pan. They'd be finished if they cracked the pan. The roar of the engine was lost in the pounding rain. Mud and gravel sprayed out behind like dragon wings.

With one hand, Chee's clung to a strap beside the door. With the other, he braced Norman against the backseat. He glanced over at Johnny, "So?"

Johnny was in his glory, spinning the wheel smoothly, gently shifting down, then hammering the gas as he powered through each turn. This was better than running moonshine.

"So, what?"

"So, where's the record?"

"In the glove box, stupid."

"I didn't think they'd fall for the old, lost-coat trick," Chee shouted above the roar of the car.

"It's all in my delicate execution."

"Johnny, I hear a siren, I think there on to us."

"Shit, that was quick."

As he spun around the corner, Beaver Creek appeared, the first crossing according to the map. Several feet of water flooded over a causeway, crashing down into a rocky gorge. They skidded to a stop.

Without a word, Chee jumped out and found a fallen aspen limb. He cleared the branches, and brace off it, as he moved carefully out on the causeway. The creek, heavy with rocks and sand, drove him downstream. He pivoted into the current and crouched low like a linebacker. Keeping his feet flat on the concrete, he side stepped, inch by inch, to the middle of the causeway. "I think we can make it," he yelled, "and Johnny, that sirens right around the bend, let's go."

Johnny revved the motor. By now, the baby blue Falcon was a mud bomb.

He dropped it into low gear and moved slowly forward, angling slightly to the upstream side of the causeway. The backend of the car started to slide as they got to the middle, and then they stalled.

"Hang on outlaws!" Johnny yelled.

In one desperate motion, he revved the engine to a scream, snapped the wheel hard left and gunned it. The frontend shot to the left and teetered on the edge. In a flash, he straightened the wheel and the Falcon lurched forward. They were moving. Several inches of water ran under the door and washed across the floorboard. As they got to the other side, Johnny cracked his door and let the water drain out. Then, they were climbing again, up through the spruce and aspen and the driving rain. Every gulch and gully was overflowing. The road-cuts were the worst, mud oozed slowly down the bank like lava, crushing everything in its path. Rocks pelted the car, as Johnny swerved around the boulders.

Chee looked back down at the causeway, "Oh no! It's the Dolores County Sheriff, with red lights flashing, siren screaming. But he's stopped. He's testing the crossing. Ha! Ha! He may turn back, but you know damn well he will radio ahead to the Naturita Marshal if they got one."

Norman started choking and coughing up blood.

"We better stop, I think it's over," Chee shouted.

Chee got out and opened the back door. Norman's eyes were glazed over. He pulled himself up and peered out the window, trying to focus. "Where are we? Why did we stop?"

"You can rest now," Chee whispered.

Norman dragged himself back from the void, "Where the hell are we?"

"We're about an hour from Uravan, but it's over."

Norman shook his head, "No, no we must go on. We only stop for one thing. I gotta go."

A sliver of a smile across Chee's haggard face, "I know, I know, give me your arm." Chee helped Norman to the side of the road.

Johnny turned to Chee, "We can stop. It's your call."

Chee checked his watch, as he unfolded the map, "It's four o'clock. I don't think we can make it, and I don't want to put him through anymore."

The rain continued to pound the roof of the Falcon, flushing mud down the front window and off the hood. The wind was coming in gusts, doubling over the smaller trees.

Norman climbed back into the car and stared at Chee, "Let's go!"

"Father, you've done enough. The spirits will understand."

"Like hell," Norman snarled. "We must destroy the bad wind. The land must be cured. Let's go."

Chee shrugged and turned to Johnny, "Let's go! We've got an hour to get to Uravan and we're almost out of gas."

Johnny revved the engine and sped off, "There's a gas station in Naturita with a phone booth. I need to check in with Maggie, see if she got a hold of Michael."

The road got better as they headed down into the foothills, but the rain kept pounding. Johnny relaxed a bit; maybe they would make it. He tried to understand what Norman had said, 'When the land is sick, the people are sick.' That part, he agreed with. But the part about getting his record to the PHS, to help cure the land, now that, was damn clever. The Navajos were way ahead of the white man when it came to spirits. As he rounded the next curve, his heart sank. A wall of mud and boulders blocked the road. He jammed on the breaks. "Son of a bitch."

Chee jumped out, "Let me have a look.

Johnny had backed the car around when Chee returned.

"Where you going?"

"It's over."

"No way. You gotta shovel in the trunk?"

"No, but I gotta an axe and a pint in the wheel well."

Johnny returned with an axe in one hand and a pint of No-Name whiskey in the other. Chee waved off the drink.

"Now, I ain't as familiar with the spirit world as redskins, but I'd say this mudslide here, is a sign. Maybe the spirit is telling us we've done enough."

Chee ran his finger down the blade of the axe. "And you're the same white boy that said you'd walk on fire, if your father asked you to? Is this another broken promise?"

"Well, ya got me there, big fellow," Johnny conceded. "What's the plan?"

"Look, I don't understand all of this either, but we've gotta finish it. Let me see that bottle," Chee took a swig. "See if you can roll those boulders off the side. I'll be back in a minute and don't let Norman see that bottle."

Johnny waded off into the mud, as Chee vanished into the forest. Moments later he returned, dragging a young, trimmed-out spruce trunk about ten inches in diameter. Johnny stood knee deep in the mud, struggling to roll a bolder off the edge.

"What's that for?"

"Old Navajo trick."

He dropped the spruce pole and surveyed the mud slide. It ran forty feet down the road. Up against the road cut, the mud was six feet high. It sloped down to the road edge, where it fell several hundred feet down the canyon.

Chee hunkered next to the car, "Our best bet is to get out there on the edge. You'll be at a hell of an angle, with one wheel over the edge, but it should work. We'll put Norman under the tree there. I'll push."

Johnny looked over the edge and cringed, "Nice plan you got there."

"If you start to roll, get out quick."

Chee propped Norman against a tree trunk and got behind the car, "Ok, let's go. Slow and steady."

Within ten feet, they bogged down. Johnny rocked the car back and forth, as Chee put his shoulder to the rear end, disappearing in a shower of mud. After several tries, they stopped. The front end was clear, but the oil pan had plowed a furrow through the mud. The backend was locked between boulders. Chee walked ahead to check things out.

"You've got another foot or so on the edge, but no further."

"You want to switch?"

Chee smiled as he wiped the mud from his face. "Hang on, let me get the pole," He jammed one end under the back axle, driving it deep in the mud. The middle of the pole rested on his shoulder.

"Keep an eye on me. I am going to work my way out to the end of the pole. When I get there, ease out the clutch. Don't pop it."

"Are you hearing that siren again?" Johnny asked.

"Yeah, he must have made it across the causeway. Come on let's go."

Chee was a load, well over six feet. If the laws of physics held true, they had a chance. He pulled off his belt and tied one end around his left hand, slipped the loose end over the top of the pry pole and tied it off to his right hand. Then he slid the belt up the pole until his feet were off the ground. He hauled down hard, nothing. This could be a great lever, or if things went wrong, he could be slingshot into Uravan. He worked the belt further up the pole and hauled down harder, nothing. He started to chant and slipped the belt to the end of the pole, and then let out a thunderous roar, and gave one final heave. The back of the car teetered up and down, then popped free of the mud.

Johnny eased out the clutch and pulled over the rocks, then gunned it wildly through the rest of the slide. He got out of the car laughing hysterically and collapsed in the mud. Then he started to sing, "Thunder, thunder he's a load; thunder, thunder, thunder road. Pretty good, huh?"

"I'd leave the singing to Roy."

Twenty minutes later, they rolled into the gas station in Naturita rapping like a tractor. They'd lost their muffler and the tail pipe was dragging. Chee pumped gas then pulled the car behind the station out of site. Johnny ran to the phone booth. He checked his watch. They were running late. It was 5:30 and it would take another fifteen minutes to get to Uravan. He shoved a handful of coins in the slot and called Maggie at the Argentine, "Hello darling, Johnny here in Naturita."

"Hold on a minute, let me shut the door. "Now, what are you boys up to?" Maggie fumed.

"Just trying to get to Uravan."

"You didn't say, you stole Norman's record."

Johnny chuckled, "We borrowed the records."

"Doctor Chavez just called. Mad as hell. He called the sheriff."

"Don't I know it. We got Norman in the back seat spitting up blood, a wind storm, a mud slide, and all kinds of evil spirits buzzing around, and Bad-Bob Johnson, the Dolores County Sheriff hot on our tail."

"You bragging or complaining?"

The wind rocked the booth, flipping the pages of the phone book.

"It's been wild, scary though. But, we aim to do what's right for Norman."

"That may be true, but you're also breaking the law and giving Mr. Stanley T. Pritchard, one more reason to send your ass a packing."

"Come on Maggie, what the hell."

Maggie ran her fingers through her hair. "I don't know, let me think. We do have one thing going for us. Doctor Chavez knows my brother, Michael, and knows he's trying to clean up the mines. Let me call him back. See if he can get in touch with Chavez."

"Hell yes. Have him call the Sheriff. Tell him there was a mistake, that they found the record."

"Actually, I'd rather have him arrest you, you miserable, record-stealing, road-running outlaw. But I'll see what I can do. Now Michael's angry as well. He's waiting for you boys, but he says do not, I repeat, do not go inside the mill looking for him. He'll wait out front in his truck. Now, go son, go."

"Aren't you the sweet thing? Next to Polly, you're my main girl."

"Yeah, see you in jail. And Johnny, while I am at it, you seem to have 20-20 vision when it comes to Norman and this uranium thing, but you're blind as hell when it comes to the acid plant."

"Yeah, well, maybe. Look, we gotta go. Norman ain't doing so well."

"What's that? Is that a police siren I am hearing? They got you?"

"The Naturita Marshal just flew by lights a flashing, must have been going eighty, heading down the Disappointment Valley road. Hee, hee hee! They missed us. This is like the Keystone Cops. Got to go."

Johnny raced off down the road. Norman was moaning and coughing blood again. The wind continued to howl, slamming the Falcon from side to side. At 5:45 pm they drove into Uravan. The tail pipe kicked sparks on the paved patches of Main Street. Chee had nodded off, but jerked awake as the mill came into view. Old memories of Uravan flashed before him. He was eleven the first time he'd seen the mill from the back of his father's pickup. At twelve, he started working in the Shamrock Mine, switching off on the big jackhammer.

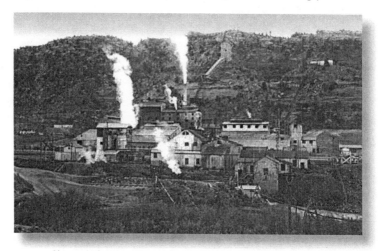

Shamrock Mine above U.S. Vanadium Mill in Uravan

Uravan was a "Yellowcake Town," the industry's term for a mill where uranium was processed. It wasn't really a town or a community. It was an industrial fortress with a boarding house. The mill sat on the west bank of the San Miguel River four miles above the confluence of the Dolores River. Cliff bands cut across the red sandstone mesas. Patches of juniper and prairie sage mottled the parched foothills. The mill slopped a half a mile or so down a sharp incline, a bastion of corrugated sheds and smoke stacks. Up above, a tanker from Rico Argentine pumped sulfuric acid into a holding pond. The caustic haze floated down into town, snapping heads and scorching lungs. Tailings were piled here and there across the river and the settling ponds cooked-off below them.

From 1898 to 1923 carnotite was mined for radium, used to treat tumors and later in luminous watches. Both caused more harm than good. In the 1930s, engineers discovered that ferro-vanadium, a steel hardening alloy, critically important to the weapons industry, could be extracted from carontite. Radium mills were converted to vanadium mills across the Western Slope. Then in 1942, the government created the Manhattan Project, a cabal to make the first atomic bomb. As it happened, the tailings piles outside the vanadium mills contained a by-product called uranium. The mills were again modified to process the tailings and the ore from the hundreds of uranium mines that sprung up. Uravan, a contraction of uranium and vanadium, grew to more than one thousand residents during the 1950s, one of several guarded colonial enclaves.

Johnny came to a stop behind Michael's truck, a paneled Public Health Service van with, Atomic Energy Commission, stenciled on the door. Norman groaned and pulled himself upright. The first thing he saw was a red-headed, Irishman peering in the window.

"The devil has come," Norman screamed.

"No, Father, that's Michael. We've done it," Chee shouted, "It's Michael, the PHS man."

Michael had an owlish face behind heavy black framed glasses. His hair did as it pleased. A rumpled khaki uniform, with a Public Health Service patch on the sleeve, covered his lanky frame. He looked perplexed, caught somewhere between anger and compassion. Johnny and Chee, caked in layers of mud, moved stiffly around the front of the car. The chrome grill was packed with mud. The radiator hissed and boiled.

Chee handed Norman his record, "Here Father, give it to Michael. He will do battle with the bad wind. He will light the fire on the mountain."

Norman pulled himself out of the car and looked to the west. The storm had tailed off. The piñon pines stood quietly in the fading light. His hair tossed in the light wind. Like an exhausted runner extending a baton, Norman handed the record to Michael.

They embraced and Norman collapsed back into the car. Michael took a moment to page through the record, and then spoke. "You had better get along. I believe the sheriff is on his way."

"We've had them coming and going." Johnny chucked. "So, what do you think? Can you use any of that? Are you in?"

"You fellows have dealt me *in* to a dangerous game. We could all end up in jail."

Chee gestured toward the uranium mill, "That don't mean, we just let *this* happen."

"Right now, I've got to get the record back to Doctor Chavez and slip the sheriff on the way down. This outlawing of yours is far from over." He stared down at them, lost in his thoughts, and then winked mischievously, "I'd like to strangle the both of you. But, yes, I am in."

"Say Michael," Johnny offered, "you may want to try the main road through Dove Creek. It might be open by now. There are a lot of cops on the Disappointment Valley Road."

After wiring up the tail pipe, they headed back to Rico by way of Norwood. It was a long, dark ride, but the road was open. Somehow they had avoided the notorious Bad-Bob Johnson, the Dolores County Sheriff. He probably hadn't made it through the mud slide.

As they headed up the grade above Ames, Chee turned to Johnny, "I'm *in*," then dozed off.

Norman bolted upright somewhere near Lizard Head Pass and began to chant. This was cut short by a coughing spell. The headlights of an oncoming car flashed across his ravished face. He smiled, "The bad wind sleeps. The fire on the mountain burns. The rest is *in* your hands."

Chapter 21

Fire on the Mountain

A rogue wind gusted over the ridge, slamming Johnny against a boulder. He fell to his knees, sucking for air. Trees were bent double. Leaves and branches sailed off. His hands had gone numb. The axe fell.

So, this was it, he growled. Now, it was up to him to keep the fires burning. He pounded the feeling back into his hands, grabbed the axe, and struggled ahead. Suddenly the sky went green, a green incandescent fog. Flashing red eyes burned through the haze. Then it took shape. A monstrous green elk soared above the treetops, snorting fire. It reared back and bugled, a thunderous apocalyptic bugle. Trees snapped. Boulders crashed down the mountain. The world was ending. So this was the day of reckoning Johnny thought. He hung under an outcrop, as the toxic blast blistered through him. So this was the bad wind seeking its revenge.

He wouldn't stop. He ducked and dodged up the mountain, timing his moves between the blasts. As he moved into the big trees, he heard a faint drumbeat. It grew louder, then a low, plaintive chant, cut through it, "Faster, Johnny, faster, the fire is dying." It was Norman.

For a moment it was calm, and then the wind returned, pounding in savage combinations, a slam from behind sent him rolling forward, a blast to the chest knocked him back. Staggering forward, he thought, get up, keep moving.

Numb with exhaustion, he staggered up and charged ahead. Fueled by a rage he dimly understood, a vague sense that he was doing right. He would keep his promise. That was it. He would keep his word.

As he moved above the tree line, he searched for the fire on the mountain. There was nothing, only bitter cold darkness. He pushed on. Strangely, the green elk had vanished, and the wind had died down. Was this too a sign? Had he escaped the bad wind? Could he still keep his promise? He lurched forward and found the cold ashes of the fire on the mountain.

He fell to his knees, gasping for breath, spewing out the bad wind, sucking in the clean mountain air. Then he was at it, a blur of axe and wood, swinging and chopping, splitting and stacking, desperate moments of dying sparks and blinding smoke, a glow, and then a flame. He rested in triumph.

Then cord, on cord was stacked around the fire. Setting the axe down, he rocked back and forth, letting the heat fill him. He had done the right thing. Rise up Lazarus! Rise up!

And then the wind returned, whirling silently around the fire. Cinders and ash spun into the sky as the wind gained speed. Suddenly, it was a red-winged serpent, wreathed in flames, spewing yellow venom. Johnny was sucked into the maelstrom as it shrieked, "It is man that has scarred the mountains with his greed. I am here to poison man."

With cyclonic fury, the serpent swept the fire off the mountain, a trail of embers, darkness.

Johnny dug himself up from the dirt and roared back, "No, you are wrong. Norman wants to cure the land and I will keep the fires burning."

He swung the axe like a mighty scimitar, faster and faster, and then he charged. Round they flew, a deadly dance of fang and axe, cutting and slashing, in and out of yellow venomous fog. Moments later Johnny stumbled out of the fray. His right hand was gone, and blood sprayed like a hot spring from his forearm. His screams were lost in the wail of the wind. Then it was quiet. The serpent had gone. He pulled off his shirt and tied it around his forearm, then dug into the ash and found a spark, a dim glow. Feeding it with bark and chips, he blew it back to life, piled on logs, and collapsed.

Seconds later he snapped awake, freezing and losing blood. To rest now was to die. He staggered forward. I've done it he thought. I've kept the fire burning. As he moved down into the big trees, he heard the drumbeat again.

"Faster, Johnny, faster, the fire in the valley is dying."

The drumbeat grew louder and louder.

Johnny covered his ears and raced down the mountain. And then he was falling and screaming, "Help, I can't do this, anymore."

Polly rolled over, "Johnny wake up, it's three A.M."

"I need help!"

"Johnny, it's me, it's Polly, it's a dream."

Johnny looked up sheepishly, "Son of a bitch."

"Was it the fires?"

"Yeah, but with evil spirits this time."

"That's the third night in a row. It's not right."

"I have to keep them both burning."

"Can you choose?"

"I don't know. "Norman said, "The fire in the meadow feeds your body, and the fire on the mountain feeds your soul.""

"So?"

"So, I don't know. It's all tied up with the acid plant and Norman's cancer. It's an Indian thing. The fire in the meadow is just getting along, the normal stuff: job, family, bills."

"Aren't you the poet?"

"You know what I mean. And the fire on the mountain is doing what's right, saving your soul and all of that. Nobody gets it right all the time."

"Maybe not, but we ought to try."

"You think it's a sign?" Johnny asked.

Polly turned off the light and threw her arm over him, "I do."

She bolted upright moments later and turned on the light, "You know, this is the first time I've ever heard you ask for help. That too, has got to be a sign.

"Good night, dear."

Chapter 22

Pray for us Sinners

**Immaculate Heart of Mary
Rico Catholic Church, 5:30 AM**

A scrim of ice coated the windows below the Stations of the Cross. Maggie knelt at the altar under a black veil, rocking back and forth, her breath floated up like smoke signals. She'd been at it a while, "Holy Mary, Mother of God, pray for us...pray for us..."

Again and again, she tried to ask Mother Mary to, 'pray for us sinners,' yet she couldn't. Normally, she was a good Christian woman, but not lately. Today, there was little forgiveness in her heart, only pain and retribution which she knew was a sin of some sort. We all sin, now and again, she figured. And most of us need Mother Mary's help. And some, she felt her heart grow cold, some like the mine bosses just needed to be shot.

"And one more thing, Mary, I know I have a foul mouth, but I'll work on it. Bless you." She crossed herself and moved to the window. Only a faint glow came from what was left of the Benally *hooghan*.

Head down and heaving, Johnny plodded up Garfield Street on his way to work. Yesterday had done him in, from Rico to Cortez, up to Uravan and then back through Norwood to Rico, the circuit. The Ford Falcon was broke and he was badly bent. But, they'd done right by Norman, and got him home. That was good.

Maggie waited inside the church entrance. It was snowing again and the rumble of the ore crusher reverberated down the hill. She stepped out as he passed. "Good morning."

Johnny's head jerked up, "Whoa! Is that you Maggie?"

"Yeah, it's me."

"What're you doing up?"

"I've had an awful night, just awful. And you're the closest dog I can kick."

"It was an awful night, and my car's a wreck."

Maggie paced back and forth on the steps, tossing her head up and down, fumbling with her rosary. "That's just tough shit, isn't it?" she stopped, crossed herself, and started over. "That's a shame, isn't it? Michael saved your ass last night. Eight hours of hairpins, pot holes and mud slides. He got into Cortez at four this morning, hand delivered Norman's record to Doctor Chavez."

"Stealing the record and all, it ain't easy to explain. And it don't sound like you're in a mood to hear about it."

"No, I ain't, you hillbilly."

"All I can say is... I don't know. Norman had this vision, and he wasn't doing well, and we got the record to Michael and we got him home again."

"Damn Johnny, don't you get it?" Maggie pounded on his chest. "Norman's dead. He was dead when you carried him in last night."

Johnny saw her face for the first time. It was a mess, yesterday's make-up and today's grief, running together. He gathered her in. For the longest time, they stood in silence, as the snow fell on them. Then, she stepped back.

"Norman's gone, and so is the *hooghan*."

"No way?"

"Chee knocked on our door at two this morning. Said Norman had passed. Said he wrapped him in a canvas and put him in the front seat of his pickup. Said the spirits of the dead were flying around the *hooghan*, had to burn it."

"Them spirits were in my house too, a flying green elk and a flaming red serpent, very strange. Navajos got themself a strange religion."

"Shut up Johnny, just shut up. That's their way. Why do you think they like you?"

"First of all, I am a fine fellow," Johnny winked, trying to cheer her up.

"No, it's because you're crazy. You're one strange fellow. They like that about you."

"Hell, I know that. You should've seen Big Thunder roll. Ripped a tree out of the ground and jacked the Falcon out of an ocean of mud. He's a Navajo warrior, amazing stuff."

Maggie grabbed Johnny around the arm and led him up the road. "Here's the deal, Crazy one. You need to take Chee down to Ertel's Funeral Home in Cortez.

"Where am I supposed to sit?"

"That's your problem, put Norman in the middle. Now listen up, this is important. Michael is going to call Chavez this morning, request an autopsy."

"What for?"

"Cause, cause his cancer probably came from working in the mines and it might help prove it. So after you drop Norman off at Ertel's, you need to meet with Michael and Chavez. Chee needs to sign a request for the autopsy."

Johnny's eyes lit up as the pieces of the plan came together. "All right, I'll do it. But, I am not going to have a job, if I don't get back to work pretty soon."

"I'll have a word with Chester. He likes you for some reason."

As they headed up to the smoldering ruins of Norman's *hooghan*, Johnny stopped. "Hey Maggie, do Navajos have a heaven?"

"I don't know?"

"...If they do...I bet Norman's up there chopping wood."

Rio Grande & Southern locomotive #419 coming up from Rico

Chapter 23

THE DANCE

At 2:30 AM, Brad drove the Big Job Pumper up to Norman's *hooghan* without sounding the alarm. It was too late to save it, and he knew he shouldn't, but he didn't want it to spread. He gazed blankly into the dying embers, trying to sort things out. Norman was a great friend. Brad would miss him. Yet, he still didn't like the business of stealing the medical record, putting Michael's job at risk with the Public Health Service. He was also baffled at Johnny, one moment the rogue warrior, the next a brooding Buddha. Some of this must have come from the coal camps. He just couldn't understand why Johnny hadn't shared his fire plan at the town hall meeting. It would have turned the tide.

Johnny and Maggie appeared, as he rolled up the fire hose. No one was in much of a talking mood. Brad needed to brief his crew at the highway department barn. Yesterday, a tractor–trailer had gone off Montelores Hill four miles south of town. Today, it needed plowing and sanding. The road was well posted from Scotch Creek on down, warning drivers to slow down, icy curve ahead. Trouble was, even on a cold day, the sun melted the snow and ice above the hill. People sped up on the dry road and never heeded the signs. Even the locals forgot that Montelores Hill never got the sun.

Brad asked Johnny and Chee to meet him at the fire house before they left town, then he headed for the highway barn. Thirty minutes later, he backed the pumper into the firehouse drive way and waited. He'd sent the Caterpillar and two dump trucks to plow and sand the road south of town. Across the street, an eighteen wheel Kenworth tanker shuddered to a stop in front of the cafe. Tankers ran around the clock, switching drivers in the morning and evening at the cafe. After breakfast, the new driver pumped twenty-five tons of sulfuric acid into the tanker and set off on the treacherous, four thousand foot descent to Uravan. Last week, they'd lost a tanker near the bottom of Key Stone Hill, nine miles west of Telluride. It hit a slick, jack-knifed, and rolled over, dumping its load in the San Miguel River. Brad rubbed his eyes. It had been a long, hard night. He nodded off worrying about clearing the roads.

Getting in and out of Rico had never been easy. There were no routes that weren't vertical and dangerous. Before the silver boom in the 1890s, the Utes and the mountain goats had it to themselves. Mining was always dangerous, but getting miners in, and getting ore out, was equally dangerous. Mother Nature seemed to enjoy hiding her bounty in the high, rugged remote.

In 1880, if you got around at all, it was by burro pack train. That year, mountains of high grade ore, assayed at over twenty thousand dollars a ton, were piled high outside the Johnny Bull Mine, waiting for the first burro train of spring. Then in 1881, the Pioneer Stage Line completed a precarious toll road from Durango, over Hermosa Park, and down along Scotch Creek into Rico. It was a bruising, three-day journey, over knife-edge peaks on a buckboard wagon. The teamsters were fearless, yet even the best of them lost a good deal of their freight over the side.

As he waited for Johnny and Chee, Brad surrendered to the smell of bacon and biscuits drifting over from the cafe. When he got back, he got to thinking about the paper he was writing for the historical society. Folks in town considered Brad a Renaissance man, equal parts philosopher and scientist. He was a busy fellow: Lead Worker for the highway department, fire chief, a dedicated conservationist, a member of the historical society, and the enduring husband of Mary Margaret Fagan, an unbridled force of nature. His paper dealt with transportation in the San Juan Mountains: the Rio Grande Southern (RGS) Railroad that arrived in Rico in 1891 and the roads that followed. He unfolded a map of the narrow gauge railroad.

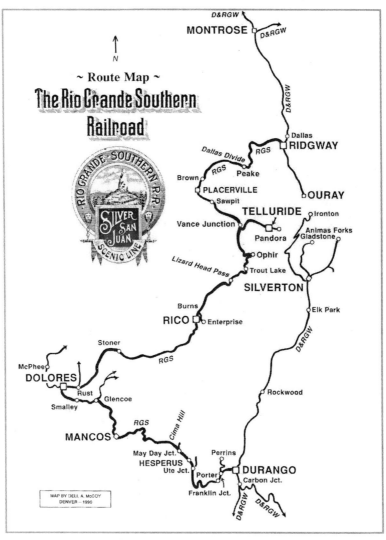

RGS map, Ridgeway to Durango

The RGS line had been awesome and terrifying, a roller-coaster ride through one hundred and sixty two miles of the most remote and beautiful country in the Rockies, crossing over one hundred and forty-two bridges. Departing daily from the Ridgeway Depot at 6,988 feet, the passenger train heaved up and over the Dallas Divide at 8,983 feet, belching steam and cinders all the way. The Pullman Cars provided for the latest in comfortable lounging and dining. It then wound its way down Leopard Creek, clickety-clack, clickety-clack, and pulled into the Placerville Depot at mile 26, at 7,316 feet. After dropping off passengers and topping off the water tanks, it climbed up to Vance Junction near Telluride at mile 38. Here it started, chug, chug, chuging, a hard pull along the South Fork of the San Miguel River, running below Ames Wall, a three-mile section of towering cliffs. No one knew when the unpredictable forces of erosion and gravity would let go a rock fall. From there, it rattled and roared its way around the Ophir Loop, a spectacular, two-mile switchback, hanging off the side of Yellow Mountain at mile 44-46, at 9,678 feet. The Loop was composed of eight, elegantly engineered bridges. Butterfly Bridge was the crown jewel, an enormous curved trestle built on a three and a half percent grade, suspended fifteen hundred feet above the roaring San Miguel River. Bridges, of course, span rivers, and rivers in the high country have a nasty habit of flooding in the spring.

Passing through the Matterhorn mining camp below Trout Lake, the engine gathered a full head of steam, and roared to the top of

Lizard Head Sheep Train

Lizard Head Pass. A two-mile stretch of open meadow, the pass was highest point on the line, rising to 10,250 feet. It was a favorite summer grazing area for sheep ranchers from 1900 to 1951. It also got more snow than anywhere on the line. In the fall, as the aspen turned to gold, thousands of sheep were shipped to market along the line. This was called a Stock Rush. Stock trains usually pulled a dozen or more double-deck stock cars.

When the passenger train left Lizard Head Pass, it gathered steam and sailed down through Rico at mile 66, at 8,827 feet, and wove its way over a series of bridges that crisscrossed the Dolores River. Upon reaching the Dolores Depot, it turned sharply southeast, through the Mancos foothills and on down to the Durango Smelter and the terminus at mile 162, at 6,512 feet.

Brad checked his watch. The boys were running late. He laid the map aside and went back to fill the water tank on the pumper, which he had forgotten to do. When he had topped off the tank, he returned to his report. Yes, the RGS was an engineering marvel and a business bonanza. Shipping ore by burro and wagon had cost eighty dollars a ton; by train it dropped to eight. The production of gold and silver doubled the year after the line was completed, yet progress and permanence are fragile notions in the high country. Geologists tell us that tall mountains become small mountains over time. That's a lot of rock fall. It doesn't pay to get in their way.

Brad had his own ideas on nearly everything. He described history as a dance between the mysterious forces of nature and man's struggle to survive. On occasion, man and nature were in step, unlocking nature's bounty and man's goodness. But there was the dark side, when greed spoiled the dance: Cattlemen overgrazing the land, miners polluting the mountains, farmers damming streams and laying claim to the water, hunters killing for hides, and leaving the rest to rot, timber tycoons clear-cutting the forest.

Brad finished his coffee and grabbed his notepad. If Johnny and Chee didn't come soon, he'd go looking for them. He was a compulsive scribbler, little happened during the day that wasn't recorded. He smiled grimly as he turned to the outline for his history report. The giving and taking of nature was as much of a mystery as the economic cycles. *Nature giveth and the market taketh*, he wrote in the margin. Perhaps this was a variation on the *Book of Job*, he'd check. When it worked, fortunes were made, as they were when a rich vein of silver was discovered deep in the Enterprise Mine. Nature, it seems, was in a giving mood. Several years later, the U.S. Treasury passed the Sherman Silver Act and silver production quickly went from one hundred thousand ounces to six hundred thousand ounces. The silver boom was on. Nature giveth and the market taketh, and fortunes were made. Was the giving and taking preordained? He wondered? Was it an endless parade of progress? Or, was it all a cockeyed coincidence, two unpredictable variables interacting in random patterns? Three years later, Congress repealed the Sherman Silver Act and the boom was a bust.

Brad made a quick note; *the effort to open up the high country was a random pattern of giving and taking.* Except for several years of prosperity, the RGS railroad had a hard time of it, in the high country. Heavy rains during September of 1909 caused the Trout Lake Dam to fail and the San Miguel River washed out miles of track. Most of the bridges and track between Lizard Head and Dolores were washed out in1911. Then in June 1927, approximately fifty men worked all night rip-rapping the banks of the flooding Silver Creek. Again, most of the bridges in Rico and below were washed out.

The first auto truck of mail was delivered to Rico on the Fourth of July, 1927. Though still treacherous, roads were being built. In April 1929, the Ames mudslide cut the RGS line in two and threatened to permanently end the freight business from Telluride to Durango. It was finally cleared in December, and so it went. Then, in the winter of 1948-49 Lizard Head Pass was closed from January to May. Neither the flange nor the rotary snowplow could cut a path in the drifts. Finally, in June of 1951, the Idarado Mining Company in Telluride switched to trucks and by November the Rico Argentine had as well.

The last train pulled out of Rico on November 29, 1951. The next day the rotary plow stood idle in the engine house. The rails at mile sixty-three near Burns disappeared under heavy snow. No one heard the cold wind moan through the trestles on Butterfly Bridge at mile forty-five. The final stretch of the Colorado State Highway145 between Dolores and Rico was paved in 1967, cutting across the graveyard south of town. It had taken forty years to complete. The dance continued a random pattern of giving and taking. Brad smiled to himself, yet you would be a fool to underestimate the power, the beauty and the wonderful mystery of nature.

He checked his watch and made a note in his folder, as the first ore truck of the day rumbled through town, shifting and braking to slow down. This was good. The pass was open. Chee's forty-six Chevy pickup followed behind the ore truck. Beyond the glittering chrome trim that adorned the grille and front fenders, it was a rust bucket. A tarp was stretched over the back. They came to a stop in front of the firehouse. Chee stared stoically ahead, hunched over the steering wheel. Johnny sat on the passenger side. The cords in his neck were flared. Both their hands were cut up from putting on the chains.

Brad stepped to the curb, "Where's Norman?"

"We couldn't shift with Norman in the middle. Put him in the back."Johnny growled. "What'd you want?"

"We've got a lot to talk about. You two need to be a whole lot smarter. First off, the sheriff is waiting for you down by the cemetery. He'll be on your tail all the way down the mountain. You go one mile over the speed limit, and he'll throw you both in jail."

Johnny snapped. "Brad, you are a smart fellow, but no one cares how smart you are in a street fight. We wouldn't be having this discussion, if Chee and I hadn't busted our ass getting the record to Michael."

"Well it won't matter much, if we don't get this part right. So... are we clear on the autopsy?"

"We've got a problem there."Johnny replied. "Chee says it's not right to look at a dead body. Says it angers the spirit of the dead, says you lose *hozho*."

Brad moved around to Chee's side, "Come on, let's stretch our legs."

They ambled down Glasgow Avenue.

"Can you tell me what Norman said in the hospital?" Brad asked.

"Said he wanted us to help him die well. Said he wanted us to help him fight the bad wind."

They walked on in silence, kicking through the crusted snow. When they got to Lucy's, they turned and headed back

"I don't agree with stealing the records and the sheriff's not done with you."

The acid tanker pulled away from the cafe and headed for the plant.

Chee wrestled with the part of him that was Navajo, and the part of him that was *bilagaana*. Did the evil spirits kill his father, or was it the bad wind from the mine? When the land is sick, the people are sick, his father's words echoed in his mind.

Brad too, struggled to bring the pieces together. "The autopsy will..." he faltered and started over. "The bad wind still poisons Norman's body. He is a prisoner. He can't go to the underworld. The autopsy...the autopsy will drive off the evil spirits, like the fire at your *hooghan*, and it could, it could give us the power we need to cure the land."

As they returned to the pickup, Chee rubbed the turquoise stone that hung from his neck. "The autopsy is a good thing then?"

"It could be, it could help us. I believe Norman would want it."

"Then, I will sign the papers."

As they passed the Burley Building, Brad glanced up at Maggie's office on the second floor. He knew she would be there. She smiled down. Brad nodded.

As Chee started to pull out, Brad rushed around to the driver's side, "One more thing, boys, after you get Norman to the funeral home in Cortez, drive on over to Saint Margaret Mary Catholic Church, at Montezuma and North Market. Michael and Doctor Chavez will be waiting for you in the parking lot. We helped build that church in 1950, you know. It's got a Baldwin organ and padded kneelers. Maggie, of course, likes the name.

Johnny put his feet up on the glove box, "Come on Brad, stop your yakking."

"Now, one last thing, Michael said that Chavez is still pretty hot about this whole thing, so tread softly. And Chee, Chavez's wife may be there. She's a Mescalero Apache, don't care much for Navajos. And, let me see. Watch out for the ice on Montelores Hill, and, of course, the sheriff."

June and Yellowman in Mountain Spring Mine, Rico

Chapter 24

THE BIG BOYS ARE IN TOWN

The mine motor with two ore cars in tow, rumbled out of the Blaine Mine portal and stopped at the mill siding. Chester, Stanley, and a Cortez lawyer crawled from the first car. Three Argentine Board members from Salt Lake City hunched over in the second. Mr. David Allen, the Chairman of the Board, was among them. They scrambled from their car, discarding hardhats, goggles and acid-proof overalls. Relieved to be above the ground, they dusted themselves off, and tried to clear the mine dust from their lungs. They were like politicians visiting the battlefield, frightened, yet hiding their fear, interpreting operational issues from a narrow legal and financial perspective, and eager to get out of town. Their gimlet-eyed focus was on productivity, corporate earnings, and stock value with little interest in the welfare of the community. The morning tour had taken them through the acid plant, and deep into the Mountain Springs pyrite mine and the Blaine zinc and copper mine.

Chester wiped the grit from his face and spit. Not a good day, but they needed to see the dirty side of the business. The cave-in at the Mountain Springs mine had just been cleared. It took down four men, one died later. Last night, the roof bolts popped on a section of the Blaine tunnel. A crew was in there now, clearing the tracks. And, at the acid plant, the new disposal line had burst and drained into the river. In spite of it all, the company was hitting most of their production targets.

The Argentine Mill and flotation pond stood just across Silver Creek.

Chester gathered the group, "One more stop, fellows."

"I think not," Stanley countered crisply. "This will be sufficient. Gentleman, thank you very much. I suggest we clean up. Lunch will be served in the boardroom at noon."

—⚹—

A formal lunch was served on the long mahogany table replete with bone china and sterling silver cutlery. Maggie refused to lend a hand. After coffee, the quarterly review meeting began. Stanley T. Pritchard stood at one end of the boardroom table, hair finely parted, suit well pressed looking like he'd stepped out of a haberdashery.

"Gentlemen, thank you all for coming. This has been quite a quarter. First off, let me say profits have never been better." The board was poker faced, giving only a slight nod of approval. No one dared crack a smile when the chairman was in the room.

"Yet, we face unprecedented challenges. Let me highlight just three. First, Union Carbide is planning to build an acid plant in Uravan. This is serious, more on that later. Second, the Public Health Service was just in town for the first time with their medical caravan, taking X-rays and testing dust levels in the mines, primarily checking for silicosis, again, more on that later. Finally, there is another governor's conference planned this spring. The Colorado, Arizona, Utah and Wyoming governors will meet with the Atomic Energy Commission and the Public Health Service. The bureaucrats don't seem to understand we're still in a Cold War. So, that is what we are up against. Chester will now present the operations report."

Chester sauntered to the front and smiled politely. His khakis were pressed for a change and his boots had been wiped clean. So far, he had been on his best behavior, but he wasn't about to gloss over the problems. He reviewed the production numbers, which were good. When he got to the capital budget, the questions started coming.

Stan Tighess, the Corporate Accountant, was the first, "Chester, the rebuild of the ore house seems high."

"It was. But, we believe the insurance company is going to cover a good deal of it."

"Let's hope so, but how do we ensure this doesn't happen again?"

"Look Stan, we've been through this before. Mining is dangerous, as you saw this morning. Some things can be prevented, and some things are just part of doing business up here."

"Surely you can prevent a fire?"

"Yes, you can. And…you can eliminate most machine breakdowns with preventative maintenance. But you can't do either, if you're playing Texas hold 'em."

Stanley blanched, "All right Chester, move on. We'll get to that later."

A slight smile crossed Chester's face, "Well, that's it for me, unless there are any other questions."

"How are you coming on the repairs to the tailing ponds?" asked David Allen, the Lead Attorney and Chairman. In his sixties and severely arthritic, he paced around the room, shoulders slumped, thumbs anchored in his vest pockets. He rarely sat. Before the Argentine, he had a brilliant career as an attorney, litigating against the government.

"We've been trying to install a disposal line that would allow us to dry stack the tailings, and…"

"Trying? Please explain." Allen snapped.

"We're running around the clock. To fully install and check the new system, we'd need to shut down for a week or so."

"So, the ponds are draining into the river?"

"Hard to say."

Allen turned to Stanley, "Is the Game and Fish taking samples from the river?"

"I hear they are, on occasion."

"Thank you Chester, you can go."

Allen circled the room on the prowl.

Stanley shuddered, as he came to a stop behind him.

"All right gentleman, let's get serious. First off, we need to talk about the acid plant Union Carbide is planning. As most of you know, they also own U.S. Vanadium, one of our biggest customers. So, we will lose that business. And as I understand the catalytic process they're using, it will severely undercut our costs. So, we may also be losing another major customer, Vanadium Company of America."

Stan Tighess nervously tapped his fingers like a pianist, "It goes deeper than that. Union Carbide has been in bed with the AEC since the Manhattan Project. They own the Union Mines Development Corporation, which did most of the uranium field research on the Colorado Plateau."

"What about that Stanley? What do you think?" Allen barked.

"Yes sir, well, we have a good relationship with the AEC, and the Cold War is still hot." He chuckled.

His humor missed the mark. Chairman Allen hobbled around the table fuming.

Bill Parma, the local attorney, jumped in, "Look, we all know the AEC wants to slow down the Public Health Service. Maybe, we can give them a hand."

Allen came to a stop behind Parma. "How could that be? The AEC is all powerful."

"The PHS has no enforcement authority, but their radon studies are starting to get a lot of attention from the press."

Allen sat and rubbed his temples, "So if we scratch the AEC's back, they'll scratch ours?"

Like a wolf on the scent of prey, Stanley's eyes flashed behind heavy lenses. "Excuse me David, but what if we took the lead in organizing a consortium of the big uranium producers, U.S. Vanadium, VCA, Kerr-McGee, along with upstream subsidiaries like Rico Argentine, and came up with a plan that restricted the PHS from further testing at our operations?"

Parma jumped up, "Then the AEC might help us out on this Union Carbide situation, like helping us construct a long-term, fixed price acid contract with the mills."

"I like the sound of that one." A wry smile flickered on Allen's face, "all in the name of national security, right boys?"

He pulled himself up and began pacing the floor, "I'll take the lead on this. Bill, you need to use your influence with the Game and Fish. Let them know we are making progress on the tailing ponds.

I'll see if I can get my contacts at the AEC to push back on them as well. Is there anything else?"

"Yes," Stanley hesitated. "They're circulating a petition down in Dolores, water contamination they say. The mayor may be able to kill it, if we sweeten the pot for his reelection."

"Do what you have to, but no paper trail."

"And one more thing," Stanley continued. "The PHS has not inspected the acid plant yet, but I think that's next on their list. Can we get some push-back from the AEC on that one?"

Allen tucked file folders in his brief case. "If the consortium thing goes well, I'll take it up with them. Okay boys, you know the drill, best of times, worst of times. And remember, loose lips sink ships."

As they were leaving, Maggie came in. "Excuse me gentlemen, Norman Benally, passed away last night. He worked in the VCA mines in Uravan for years, and his son, Chee, is a supervisor at Mountain Springs Mine. We're passing the hat, wonder if you wanted to make a contribution."

The room fell silent.

# Chapter 25

½ Inch Socket on a ½ Inch Nut

The pickup fishtailed from side to side on the curves. Light powder covered hard ice. As they passed the graveyard, a patrol car pulled behind. Johnny dozed fitfully on the passenger side. Chee held a paper cup in one hand and spun the steering wheel with the other. Coffee splashed in all directions. Pale and haggard, he willed himself awake, finally giving up on the coffee.

As they approached Montelores Hill, a notoriously unbanked grade, Chee shifted down and pumped the brake. They rounded the curve slowly and stopped near the bottom where the ore truck had gone off. It lay on its side in the river. The river banks were blanketed in snow, setting off the brown, acid-stained river rocks. The ambulance had come and gone.

The elevation drops nearly three thousand feet in the thirty-nine mile drive to Cortez. Jagged mountain peaks give way to the flat, desert emptiness. The river carves its way down the valley, crashing through narrow gorges, then runs gently across open meadows. Somewhere near where Bear Creek joins the Dolores River, things start to change. Red rock ledges, molded as if by a potter, displace tall stands of spruce and aspen.

As the road flattened out below Stoner, Johnny finally spoke. "You tied him off, did you?"

"Tried to, covered him in snow, tied down the tarp."

"He's taking the ride pretty well."

"He liked this country."

"I am probably going to say this wrong, but, I want to say, I am sorry Norman died, and I will really miss him, and you had a fine father, and he was pleased that we got his record to the PHS."

Chee nodded without expression, as stoic as an oak, and went back inside himself. They drove on in silence. Johnny was troubled. The uncertainty was eating at him. He'd shredded too many knuckles on adjustable, all-purpose wrenches. For now, he needed a snug fit, the certainty of a ½ inch socket on a ½ inch nut. His sense of right and wrong ran deep. He knew where he stood on that. He was fascinated by the Navajo Way, yet struggled to understand it. Norman had said in the hospital, 'When the land is sick, the people are sick.' At the time, it seemed no more than a wise saying, but now, Norman was dead, probably from working in the mines. Now, it had great meaning. Then there was Norman's last request, taking his record to the PHS. Shit, man that was bold. Yet it seemed like a long shot, maybe not. But the flying elk and the flaming serpent, what was that all about? And then there was Chee, one minute the raging warrior in a river of mud, the next a pillar of stone, calmly enduring whatever came his way.

Chee slowed as he entered the city limits of Dolores, but not soon enough. With the red light flashing and the siren wailing, the sheriff pulled him over and got out. He was a wedge, a donut in one hand and a ticket book in the other. His gun belt disappeared behind his stomach.

"Going a little fast there weren't you, sonny?" he shouted a little too loud, searching for an authoritative tone.

"It's posted at twenty-five, and I was going twenty-five."

"Let me see your driver's license and registration."

He stepped to the front of the pickup to check the plates then returned.

"You, Chee Benally?" he said, leaning inside the window.

"I am."

"Chevy made a nice pickup in forty-six, didn't they? That chrome grille is damn sporty. You ought to clean it up."

Chee and Johnny suppressed a smile. Everyone had heard about Bad-Bob Johnson, the Dolores County Sheriff. They said he was a strapping fellow, no one to fool with. This wasn't Bad-Bob.

"You're the sheriff, are you?" Johnny asked.

"Bob's got his hands full today. I am Luther Parson, his assistant. You got a problem with that?"

"Hey Chee, you got a problem with that? No, and I don't have a problem with that. So, no sir, Sheriff Parsons, no problem here."

Luther turned to go, then stopped, "What's under the tarp?"

"Nothing," Johnny said, looking to Chee for an answer.

"Nothing, then what's the tarp for?"

"Nothing, but snow," Chee answered.

Luther pulled back the tarp and considered. "So, why you got a tarp on it?"

Chee jumped out, pulling a picture from his wallet. "It's for my little nephew here. He lives in Mexican Hat, out in the desert. They don't see much snow out there."

The sheriff shrugged, "That's probably bullshit, but it's pretty good bullshit. All right boys, I'll be keeping an eye on you." He got into the patrol car and drove off.

Johnny got out and helped Chee tie off the tarp.

"Damn Chee, that was slick."

They looked at each other out of weary eyes and burst out laughing. They threw their heads back and laughed. They stamped their feet and laughed. They danced their way around the pickup and laughed a loud raucous, cathartic laugh.

Johnny wiped the tears from his eyes, "You got a nephew in Mexican Hat?"

"Where's Mexican Hat?"

As they drove into Dolores, Chee looked back at the tarp, "You know what Norman liked most about Dolores?"

"Going to the movies?" Johnny offered. "No, hold on, going to the movies where the cavalry got scalped."

"He liked that, but that's not it. Even more, he liked the Hollywood Bar."

Johnny looked at Chee and smiled warmly. He was back on his horse. His eyes flashed playfully, the gray stillness was gone. It was 8:30 AM. They were due at Ertel's at ten. "We got time, if it's open."

"It's always open, paleface."

Chapter 26

BREAKFAST IN HOLLYWOOD

The Hollywood Bar, with its promise of romance and glamour, was located just across from the Dolores Train Depot. They say, on a clear night, the HOLLYWOOD BAR sign, spelled out in large block letters just like the one in Hollywood, California could be seen by passengers planes flying over at thirty thousand feet. They say.

The proprietor, who was never the same, boasted that, *the beer's cold and the people are friendly*, to which the locals would add...*when they ain't fightin'*. Two long windows, cluttered with beer posters, ran beside the front door. Hand written notes were tacked to the door: a 30-30 Winchester for sale, rooms to rent, snow shovelers and wood choppers for hire, bingo on Thursday, open mike on Friday, a revival meeting on Sunday.

Johnny scanned the tattered notes, "I'll be damned, Roy's on tonight. It says: rock and roll and songs of local concern. Roy is expanding his act."

A rancid concoction of cigarette smoke and stale beer hit them as they entered. Elk heads were mounted above a long bar. A wagon wheel, trimmed in holiday lights, hung from a tin-sheeted ceiling, with the occasional bullet hole. The lights cast a rainbow of colors on the pool table. Over by the dance floor, the Wurlitzer flickered off and on, sending a distress signal. A thin angular man in yesterday's shirt and a drooping bow tie appeared from the back, pulling a mop and bucket.

"Howdy boys, name's Cal. You need an eye-opener?"

"Three shots of your best whiskey," Chee ordered. They sat down at the bar, leaving an empty stool between them.

Cal steadied himself as he moved behind the bar, his ancient joints cracked. "You got it." He filled three shot glassed with Seagram's VO.

Chee lifted his glass, "To Norman!"

They toasted the glass on the bar between them, downed theirs, and hammered the empties on the bar.

"Another round," Johnny shouted. "And pour one for yourself there, Cal.

Johnny lifted his glass, "To Norman!" Again, they toasted the glass at the empty stool, emptied theirs, and hammered the empties on the bar. Two full shot glasses now stood in front of the empty stool.

"Say fellas, none of my business, but is Norman running late?" Cal asked.

Chee paused and collected himself, "Norman…Norman Benally is my Father. He liked this place. Did you know him?"

"Oh, hell yes, I know Norman. We worked together down in the hole in Uravan. I haven't seen him for awhile. How's he doing?"

"He's doing fine. He died last night. We're taking him down to Ertel's."

"Ah Jesus, sorry to hear that," Cal slumped against the back of the bar, tried to light a cigarette and gave up on it.

"Not many of us old timers left. You must be Chee, ran the big jackhammer when you were twelve, I hear. Your daddy was mighty proud of you, young man, never doubt that."

Chee nodded and thought back to the day his father handed him the big hammer. He had to get his knee under it just to get it off the ground.

Johnny pointed at the two full shot glasses on the bar, "Chee, how you want to deal with this? We gotta' go."

"Cal, pour one more and let's go outside."

Chee left a ten on the bar. When he got to the pickup, he untied a corner of the tarp and pulled it back. Then raise his glass in the air. "To Norman!"

They emptied their glasses and Chee buried his glass in the snow that covered Norman. They did the same.

Cal shook their hands, "Thanks for including me. Say fellas, it's none of my business but when's the funeral?"

Chee looked at Johnny, "Don't know yet, next week early, maybe Monday or Tuesday."

"Now, it is none of my business, but you may want to have him embalmed if you plan on an open casket.

Chee was confused, embalming, autopsy. What did they mean? None of this was part of a traditional Navajo funeral. He said nothing.

"Thanks again, Cal. We'll keep it in mind," Johnny jumped in the driver seat and started the engine, hoping to end the conversation.

As they started to pull away, Cal stooped over by the window, "And fellas, I know you're from Rico. If you drop by on the way home, be careful what you say. Folks around here are fed up with the acid plant and all the shit they're dumping in the river. You're welcome, but I wouldn't go bragging about Rico, like some of them do."

As they headed down to Cortez, Johnny chuckled. "Say Chee, the shot glasses in the snow, is that a Navajo thing."

"Can't say. Wherever he's going, I figure he'll need a few extra shots."

Chapter 27

WAKING THE SLEEPING UTE

The country flattened out as they wound down the foothills below Dolores. Rolling grasslands with patches of lodgepole pine and scrub oak gave way to farms and fenced-in pastures, all covered in snow. Here and there alfalfa bales were flaked on pastures. Cattle moved back and forth in steaming slumber, from hay to water. The foothills marked the transition from the high country to the high desert, like the center line on a blacktop. Off to the east in the distance the snowcapped La Plata Mountains dominated the skyline.

The tension drained from Chee's face as he looked out on the reservation, *Dine Bikeyah*, Navajo Country, coming home.

Johnny looked over, "You feel like talking?"

"I don't mind."

Three shots of VO had loosened their tongues.

"I heard Brad mention *hozho*. What's that all about?"

"That's not an easy one. You need a medicine man for that."

"I ain't sick."

"And hell, that ain't funny, paleface. For me *hozho* is being at one with nature, being connected, not boxed up."

"Boxed-up?"

"*Bilagaanas* put things in boxes."

"We do?"

"You have a box for your head, and a box for your heart, a box for man, and a box for the mountains. You separate what is one."

"Chee, that's just the way it is. A tree's a tree, and a rock's a rock. I'm ok with that."

"But if you take away the boxes, there is oneness, harmony. That's *hozho* head and heart, man and mountain, all one. When you have *hozho*, you feel the wind wander through it all, giving it life. That is *nilch'I* the Holy Wind."

The town of Cortez appeared as Johnny came around the curve. "I like that. *Hozho* then, is when it's all bolted together? I have had those moments."

"Come on, Johnny, bolted together, sounds like you're repairing a car. It's better than that. For me, it's a path, a way of life, it is being in balance with the nature."

"So, how do you lose it?"

"I know some things, but much is hidden. I know, you must listen to your heart, you must learn from the land, the sunrise, the sunset, the change of seasons, the bugling elk, the roaring bear. Remember, my mother was a Lakota Sioux, so none of this is clear to me."

"But, how do you lose it?"

Chee rubbed the turquoise stone that hung from his neck and thought deeply. "Let me try. When you see all things as one, you are part of it, a common family, a common land. We have a saying, *Mitakuye Oyasin*—we are all related, the mountains and the rivers, the flowers and the trees, the fish and the birds, and all the tribes of man. But, when you put things in boxes, they are outside of you. Then, some will take, some will destroy. I don't know why, but I believe it is so.

Johnny turned off 145 onto Main Street. Several blocks later he took a right onto North Market and came to a stop in front of Ertel Funeral Home, an imposing Southwestern structure with adobe colored walls. A mission bell hung from the middle of an arched entrance. To the right viga beams protruded from a flat roof, to the left was a pitched tile roof. At the side, a gleaming black Cadillac Landau stood at the ready. As they pulled in the driveway, Chee's legs started to shake and perspiration broke out on his forehead.

"Let get out of here, I don't want Norman bombed."

"It's embalming, and I don't know what it means either," Johnny sighed.

Just then Keenan Ertel appeared. His grandfather, J. Walter, founded the funeral home in 1921. Keenan was as polished as a river rock; impeccably dressed in a dark suit and string tie.

"Good morning gentlemen, you must be Chee, and is it Johnny?"

"That us," Johnny said getting out of the pickup, "Norman's in the back."

"Well, let's go inside, my assistant will care for the body."

"You'd better hold off on that Mr. Ertel. We got a problem," Johnny stammered.

"Oh?"

"Chee here, is not very happy with *bilagaana* funerals. He doesn't understand the whole business of autopsies and embalming and neither do I. Navajos are mighty touchy about their funerals and they don't want any bombing."

Chee refused to get out of the pickup, "No autopsy, no bombing, let's go!"

Keenan walked around the pickup and spoke in a soothing tone, "Chee, we are here to serve you, to show respect and dignity for the deceased. We do need to complete a death certificate, but nothing more, unless you request it. Please come inside. We'll sit in the chapel and talk it over."

It was silent and peaceful in the chapel. Above the altar, was a stain glass window with a clipper ship sailing beneath a marbled blue sky. A rainbow of soft light showered the room. *Crossing the Bar* was inscribed at the bottom of the window. Hand carved wooden pews stood in rows on both sides of the aisle. Chee was subdued by the soft colors and the silence. It seemed to trigger good memories from the past. They sat quietly for the longest time in the back pew of the chapel.

Finally Keenan spoke, "Chee, your father was well thought of around here and we will do whatever we can to celebrate his life. Doctor Chavez called yesterday and said you were coming and requested an autopsy. That is all I know about it."

Chee turned to Johnny, "Cal said we needed a bombing, embalming whatever."

"Can you help us with that one Mr. Ertel? Johnny asked. "What is embalming."

Keenan gently responded, "First of all, there is no bombing involved. Embalming is a procedure that pumps fluid into the body, so it can be preserved for an open casket funeral."

"So Norman's blood is pumped out of his body? Chee exclaimed. "Forget it, let's go get Norman and get out of here!"

"Hold on, Chee." Johnny said. "So, Keenan, can I call you Keenan? Do we really need this embalming?"

"Please call me Keenan. There was no request for embalming and I don't believe it is necessary if you are planning the funeral on Monday."

Chee was still confused, "So, no embalming, but what about the autopsy?" Chee turned to Johnny, "Brad Fagan said an autopsy would help drive off the bad winds from the mines. Is this true?"

"Hang on." Johnny took a moment to think about what he wanted to say. Turning back to Kennan, "We've been told that Norman died of cancer. Will an autopsy help to prove it?"

"It will certainly help explain the cause of death."

Chee nodded, "Do the autopsy, skip the embalming."

Keenan, the consummate professional, began the difficult task of guiding Chee through the minefield that a funeral entailed: the cycle of grieving, the integration of Navajo burial rituals, the shaping of a fitting remembrance. The body would be prepared here for burial and then on Monday it would be taken in the hearse to the Native American Church in Red Rock, Arizona. Finally, the hearse would lead the procession to the family burial site on the Benally plot in Cove.

As Keenan completed the death certificate, Chee wandered over to the stained glass window. He was fascinated by the sailboat dissolving into the sunset, crossing the bar, a marbled blue sky, splintered with salmon rays. It was healing.

Keenan walked them out to the pickup when they finished, "I'll get this request over to Doctor Chavez and call the pathologist. It will all work out fine."

—∞—

They drove up North Market to Montezuma Avenue and pulled into the parking lot of Saint Margaret Mary Catholic Church. Michael Burns and Doctor Chavez sat in a paneled Public Health Service pickup.

They gathered near the hood of Chee's pickup. For the longest time, Michael just stood there, biting his bottom lip, reigning in his anger. His red hair was in disarray and dark shadows hung under his eyes.

"Morning boys, you know Doctor Chavez. Let me start this conversation in a civilized manner."

Chee and Johnny shuffled uncomfortably, heads down, hands in their pockets.

Michael nervously tapped the hood of the pickup, "If I thought it would do any good, I'd take a two-by-four to the both of you. So, listen up. We're in a deep hole here and we've got to be damn clever to get out."

The wind whipped through the parking lot. No one spoke for several minutes.

Dr. Chavez stepped forward, "Personally, I'm very upset. This whole business of the relationship between lung cancer and uranium radiation is very complex and very political. You boys have gone outside the system and broken the law. This can greatly weaken our case."

"The doctor is right," Michael added, "He's got the State Medical Board on his back, I've got U.S. Vanadium and the AEC on mine, and you boys have Bad-Bob Johnson and Mr. Stanley T. Pritchard on yours. That's what you got us into with your outlawing."

Johnny kicked against the front tire and inhaled deeply, "Let's get this straight. We'd never have gotten this far, staying inside the law. If the system's broke, you damn well have to go outside the law. Michael, last night you said you were in. It don't sound like it this morning."

"I am in, but I am not happy about it. This has gotten out of hand. We've got to fix the system"

"For two years now, we've been trying to do that in Rico and it's still a mess. And tell me, Mr. Public Health Service, how many miners have you tested so far?"

Michael cringed, "Too many."

Dr. Chavez intervened, "Chee, did you request the autopsy?"

"It will be done this afternoon."

"Good, I've got a release I'd like you to sign. It authorizes the pathologist to give me the autopsy report. As you know, I've treated a lot of miners. It's hard to say, but this could be very important. Beyond that, get back to work and keep quiet."

"Maggie tells me the concerned citizens of Rico are having a meeting next week," Michael said. "I'll be there. And boys, I am way out on a limb. If the PHS gets wind of this, I am done."

Doctor Chavez checked his watch, "Gentleman, there is no turning back on this. We've got to work together. If we're right about the autopsy, the big boys at the AEC, at U.S. Vanadium, and at the Argentine won't want it to go public. Let's see what the report says and go from there."

Chee pulled into the Skyline Gas Station at the intersection of highway 145 and 184 above Cortez, after finishing their shopping. They got out and stretched, relieved that the hard part of the day was behind them. After filling the tank, he returned with two cokes. They sat on a bench in front of the station and watched the sun slip behind the Ute Mountains, a singular range that rose four thousand feet above the Montezuma Valley. The sky was a cloudless blue, splintered with salmon rays. As mountain ranges go, it wasn't much, running only twelve miles from north to south. Yet it dominated the horizon. Its contour was a mix of domes, flats and peaks, formed million of years ago by igneous intrusions and the relentless pounding of wind and rain. Michelangelo, himself, couldn't have done a better job sculpting the mountains into the shape of a reclining Indian, known as Sleeping Ute.

The domed top of Marble Mountain formed the head, a feathered headdress tapered to the north. Ute Peak, the highest point in the range, formed the crossed arms of the Great Warrior God. The ribcage, knees, and toes extended down the rest of the range. The reclining figure was neither a mirage, nor a flight of fancy. It burnt into your mind, once perceived, a convincing representation, yet hauntingly iconic.

Chee and his father had watched the sun set on Sleeping Ute many times, the first when he was a boy. As a Navajo, Norman didn't always get along with Utes, but he loved this legend. All tribes could call on Sleeping Ute for help.

"Do you see it?" Chee asked.

"It's a fine sunset."

"Yes, but follow the ridgeline."

"It rambles, don't it? I'll be damned. Is that a fellow lying down?"

"That's Sleeping Ute. Long, long ago he was a Great Warrior God."

"That's pretty cool. I would have missed it. He's sleeping, is he?"

"Recovering from wounds received in a battle with the Evil Ones. The Ute Tribe say he will rise again to help against the enemy. "

"Well, let's wake him up."

"Chee smiled, "We may need to. Come on let's go."

Chapter 28

A Call To Action

It was dark when Chee and Johnny finished up in Cortez. The lights from the Hollywood Bar lured them in as they drove through Dolores. From inside the bar, *Jailhouse Rock* boomed out the door and across the parking lot. Roy and Trish Vanderville were on stage and the place was jumping:

> *Everybody in the whole cellblock*
> *Was dancing to the jailhouse rock.*
> —From Jailhouse Rock by: Jerry Leiber, MikeStoller

Leon and the rest of the cowboys from Stoner anchored one end of the bar. The locals had the other. Trish strummed her guitar as she spun around the stage like a mad gypsy, in a battered derby hat with tail feathers flying, part Bolshevik and part Indian princess. Her crinoline petticoat swirled above stripped tights and Doc Marten boots. All in black, Roy pumped his hips like a goat in rut. His hair was a marvel. Plastered in pomade, the side walls and tail were rock-solid. His jelly roll in front moved to its own beat.

Johnny sipped his beer and enjoyed the show. This had been quite a day. Roy's Elvis bit was improving, sort of, and how about Trish? Johnny had never met anyone quite like her: a debutante from the East Coast, who led protest marches in Berkeley, and now taught school in Rico. They came over when they finished to say hello and expressed their sympathies to Chee.

Roy pulled Johnny aside later, "You gotta fill me in on your mud run to Uravan. Maggie wouldn't say much."

"We'll talk later. Say Roy, you guys are great."

A group of locals at the bar started to argue.

"Then, you tell 'um!" someone shouted.

A bruiser in a pea coat growled back, "I believe I will," and made his way over to Roy.

"Hey Elvis!"

The bar fell silent.

"I hear you and Pocahontas there are from Rico. The ones that have been polluting our river?"

Roy stood up slowly, a goofy grin spilled over his face. He put his hands on Chee and Johnny's shoulders to keep them down.

"Howdy partner, name's Roy. What's yours?"

Normally, Roy would have gone into his prefight warm up, cracking his knuckles, rolling his shoulders, shaking out his arms. Tonight, he stood calm and relaxed. As a singer he was okay, as a street fighter he was a holy terror.

"Mine's Jimmer, didn't come to talk."

"Well Jimmer," he toyed with him, "it's your lucky night there bad boy. You are looking at a changed man," he looked over at Trish and smiled. "I'm not into fighting anymore, political actions my game, changin' what ain't right."

Jimmer put his face into Roy's, "Well, ain't that sweet. But that don't clean up the river, does it?"

The cords in Roy's neck started to dance about. He wasn't good at staying out of fights. Johnny and Chee stood up. Jimmer's buddies joined him. It got even quieter. A toxic blend of testosterone and estrogen flashed through the bar like dry lightening. The primal urge for combat and coupling could burn the place down. Trish jumped up on a chair, "All right gents, I got something to say. Roy's right, we're trying to change things, and for us, it starts with our music."

Jimmer, like every young stud in the bar, had taken a liking to Trish. He gave her an angelic, toothless smile, "Well little darling, I guess you better sing us a song."

Cal, the ageless bar tender, shuffled over from behind the bar, followed by his son Junior, a gangling giant, who looked like he'd escaped from a reform school. Cal pulled the plug on the Wurlitzer and turned to the crowd, "The next sum-bitch that starts a fight is eighty-sixed, right Junior?"

Junior smiled malevolently down at Jimmer, who drifted back into the bar.

Trish stepped forward, as they strapped on their guitars, "Most of what you're going to hear is new stuff, so be kind.

She tuned her guitar, then looked out on the crowd.

"Since we bombed Japan in World War II, all of us, in one way or another, have been caught up in the Cold War, especially here on the Western Slope, whether it's digging uranium, or making acid, or fattening cattle to feed the miners. Like it or not, we're all part of it."

The locals were angry about the pollution, yet patriotic. No one knew where she was headed. She glanced over at Roy and nodded, "We call this one, The Cold War Blues."

> *Breakin' the law*
> *Bringin' in the funds*
> *Breakin' the law*
> *Getting' out the tons*
>
> *Call in the lawyers*
> *Grease a few palms*
> *Just doin' business*
> *Ain't nothin' wrong*

They had a strange harmony, Roy, the raspy baritone, Trish the trilling tenor. A few people clapped along.

> *Acid burnin' in your chest*
> *Ain't no cause to take a rest*
> *Gotta keep that ore a slushin'*
> *If we're gonna beat the Russians*

By now, most were back on their feet. Those at the bar were tapping out the beat with their beer bottles.

> *Preachin' safety is just fine, but*
> *Boys we're here to work the mine*

They finished up and the crowd went wild, chanting for more. Trish stepped forward, "How about ole Elvis here, ain't he something?"

Roy made a sweeping bow, as the applause washed over him. This was a big night for a boy from the coalfields.

Trish continued, "Now singing the blues is one thing, turning them into rainbows is another. Here's another new one. The working title is, *A Call to Action*. The crowd had settled back in their seats.

Trish and Roy strummed down wildly, creating a militant beat. Then they were off, wandering through the crowd.

> *There's power here, there's power there*
> *There's power almost everywhere*
> *But, now's the time to all join hands*
> *Now's the time to make a stand*
>
> *Bitchin', cryin', whinin', groanin'*
> *Limpin', laggin', wailin', moanin'*
> *Ain't no way to cure the land*
> *Ain't no way to fight the man*

They paused. The mood had changed. The crowd was coming together, many streams flowing in the same direction, kicking up white water. They were ready to act, most of them anyway. Somewhere near the back, a table of roughnecks started to chant, "Don't bite the hand that feeds you."

They were drowned out by a chorus of, "Now's the time to take a stand."

Johnny turned to Leon, "Who are those boys?"

"That's Okie Tom Barlow and his truckers. They haul ore from Telluride and Rico."

"Assholes."

Then, they were off again. Roy boogied around the pool table, strumming wildly, showered in holiday lights. Trish was hoisted onto the bar and spun like a top, a blur of pink crinoline and yellow-soled boots. Dust drifted down from the ceiling, as whiskey bottles rattled in three-quarter time.

> *Acid rain and river rust*
> *Yellow dirt and toxic dust*

> *Mother Earth is cryin' loud*
> *Time to organize the crowd*
> *Time to organize the crowd*

The room fell silent as they finished up, then exploded with stomping and cheering. Trish jumped off the bar and returned to the stage. Roy held up his hand for silence, "Thank you, thank you, thank you. I owe it all to Trish here, she's the brains of this outfit."

Trish bowed politely.

The crowd chanted, "More, more, more."

Roy stepped forward, "They say no one has died in this Cold War, but that's not true. Last night, Norman Benally died. No one is sure why he died, but living in Rico and working in the uranium mines for years didn't help. Chee, over there, is his son. Come Monday, we're burying Norman down on the reservation."

Chee nodded politely, uncomfortable with the fuss.

Roy continued, "So, I want to end the night with two requests. First, I'd like as many of you as possible, to join us at a meeting at The People's Congregational Church in Rico, next Thursday evening to discuss the pollution of the river. And second, since Leon over there has the biggest hat in the county, I like him to pass it around for the Benally family.

Trish rejoined Roy and they sang the chorus of, A Call to Action:

> *Now's the time to all join hands*
> *Now's the time to take a stand*

And most of them did join hands and chanted the line again and again.

People's Congregational Church Rico

Chapter 29

BUILDING A COALITION

February 1964

"It is difficult to get a man to understand something when his job depends on not understanding it." —Upton Sinclair

And so they came, a few crusaders for change, a few defenders of the sacred here and now. Most were hopeful but unsure of what could be achieved by studies, petitions, and injunctions. Yet they came, a motley crew of needs and wants. Could the collective voice of the people slow the grinding gears of commerce? Could they make a stand or would they step back and quietly endure?

A group of thirty or so had assembled at the People's Congregational Church to discuss the pollution problem. It was a mix of those for and against cleaning up the plant. Polly and Johnny sat near the front with Trish and Roy, who was sporting a clean shirt. Chee and five other Navajo miners huddled in the corner conversing in Navajo. They had worked in the uranium mines in Slick Rock before coming to work for the Argentine. Michael Burns sat with Brad and Maggie Fagan with Norman's autopsy report in his pocket. He wiped the perspiration from his brow and tried to remain calm. Jim Rychtarik dozed quietly beside Bill Meyer, the mayor of Dolores. The meeting had been going over an hour and everyone had had their say. Val Brown, the librarian, rambled on for the last ten minutes. Her husband Alton worked in the Argentine Mill and refused to come.

"So that's life," Val concluded philosophically. "You take the good with the bad. That's what Alton says. Most of you love to camp, right? Well you wouldn't give up your campfire if a little smoke got in your eyes, now would you?"

A few smiles appeared like weeds in a flower patch.

Not to be outdone, Maggie jumped up, "Well, tell me Val, how long would you sit in the smoke before you moved?

Everyone smiled and most chuckled, but not enough to change their mind. They held onto their opinions like a bear trap. The more you pulled, the more they resisted.

"Maggie you've had your say, sit down, Clem's next," scolded Susan Clever, guarding the podium like a pit bull. Susan's life as a bully had started in grade school when she had been a foot taller than her age group. Now she was of average height with the curse of a childhood bully to deal with.

Clem Reed, a longtime Argentine miner, hobbled to the podium. He'd never been right since the Blackhawk cave-in. He struggled to clear his throat, then got to coughing, a harsh bronchial rattle. As he began to speak, Okie Tom Barlow and his two brothers blew in the door. They moved with a swagger, big, barrel chested men with long greasy hair. Clem was forgotten. The Barlows were roustabouts from the oil patch in Oklahoma that had moved to Dove Creek. They owned a trucking company and hauled ore for the Argentine. Okie strutted to the center of the room casting an evil eye at the gathering, "I got the marshal outside. Are you folks planning to incite a riot?"

Susan Clever went red, searching the room for help. Myron, Roy, and Johnny stood up and crossed their arms.

"Okie, you boys will get your chance to speak in a minute. Right now, Clem has the floor," Susan stammered.

Roy started popping his knuckles and whispered to Johnny, "I'll take Okie. "You got the brothers."

Johnny chuckled, "Hang on, big boy."

Roy rubbed his chin, "We huntin' or trappin'?"

"Damn Roy, that's good, where'd ya get that one?

"That's what my daddy use to say."

"Well let's trap a while longer."

"Lost your nerve there, have ya, Johnny boy?"

Pleased with their entrance, the Barlows grinned broadly and sat. By now Clem had his cough under control and began to speak, "I say we hold off. Ches-ters a gud man, been gud fur me. If he says the Burro of Health is a checking things, that's gud nuf fur me."

"Geez Clem, a burro is a beast of burden, and it's not the Bureau of Health either, it's The Public Health Service," Hartley Lee snapped.

Irene, his wife, turned to Hartley and chuckled, "Ole Clem may have it right."

Clem droned on for another five minutes, a frayed hunting cap pulled to one side of his head.

"Yur luk-un fur trouble, if yur luk-un to take on the Ar-gen-tine," he concluded and set off coughing again.

Lucy Fahrion came forward as Clem finished. She was a born peacekeeper, yet deeply committed to change. "Clem you been mighty loyal to the Argentine over the years, that's no secret. We all want this to work out with the plant, Lord knows we need them. We've just got to find a way."

She wouldn't take the lead, but she would follow a well-reasoned plan. The Navajo miners got louder and louder. Thunder was doing his best to translate and settle arguments. It was time for a break.

Susan Clever stepped to the podium. "All right, let's take ten. Then let's get back and decide what we want to do."

Everyone headed for the door.

Johnny sat next to Polly on the porch steps, "I am all for a good fight, but changing the way people think is hard work."

Polly smiled, "Spoken like a true, low-life, street fighter. Just hang in there, the fact that we got this kind of a turnout is a good sign."

Off to the back beyond the porch lights, the Navajos gathered next to Chee's pickup. Russell Yellowhorse was arguing in Navajo, then shifted to English, "The Chief made a deal with Pritchard, no union, no trouble."

"No English," another shouted. Russell looked around the yard, paused and went back to Navajo. Several others, weary of all the talk, wandered off. Chee patiently corralled them and finished the powwow.

Michael Burns walked up as Chee concluded. "Sorry I couldn't catch you earlier. I've got the results of the autopsy."

"What did they find?"

"Norman died of oat cell carcinoma that had spread from his lungs to his brain."

"So, it was the bad winds from the mines?"

Michael looked away, as his voice cracked, "That's been the pattern."

"The pattern, what kind of word is that?...the pattern," Chee asked.

Guilt burnt through Michael, "A scientific word that hides the truth."

Chee looked over at the crowd outside the church, "You gonna tell them?"

"Yes, I suppose it's high time." Michael's head was spinning. This was the biggest crossroad of his life. An exacting scientist, he had carefully gathered medical information from miners for years. He had spoken-up to the PHS and to the AEC on his findings. He had done everything he could to work within the system, and still had hit the wall. There was no turning back. After talking with Chee, Michael met with Brad and Susan Clever and they agreed to let him speak.

Susan made the introduction when they regrouped. Michael took a deep breath and stepped to the podium. He held on to steady himself, "Good evening, as some of you know, I have worked for the Public Health Service since 1940." He paused, struggling to gain his composure.

"Tonight...well, tonight...I will be putting my job on the line for saying what I am going to say, but to be honest I've reached the point where that's no longer important to me. It's time," his voice broke, "it's time I speak up."

Dozing heads snapped back. The room sizzled with tension like a high peak before a storm.

"What I want to say is...the Atomic Energy Commission has handcuffed the Public Health Service. We're no longer in the health and safety business, we're tied in red tape and wrapped in the flag. It's all about national security and the arms race, and if you can believe it, "safe" atomic energy. Now this is big. Let me give you a few facts, eighty percent of the uranium purchased by the Manhattan Project came from Uravan and the Durango mills, and from 1948 through 1960, Colorado produced uranium ore valued at over 133 million dollars."

This poured out of Michael like a draining abscess. Chee translated nonstop, struggling for ways to explain words like, 'red tape', 'national security', and 'arms race'. Michael turned to a tab in his binder, and calmed himself and the exacting scientist took over, "There are many health hazards in the uranium mines and mills, and in acid plants like here in Rico that produce sulfuric acid for extracting uranium. We've known about the radioactive hazards for a long time and have done very little."

Maggie was beaming, she whispered to Brad, "I saw this coming, but I never thought it would happen tonight."

Michael stood tall and continued, "Now, I want to talk about Tom Van Arsdale, a uranium miner we've tested over the years. He worked for U.S. Vanadium in Uravan and died nearly four years ago. Tom had regular X-rays beginning in 1940, came down with silicosis of the lungs in 1953, and died of lung cancer in 1956, technically of oat cell carcinoma. That's sixteen years after his first exposure to uranium."

Clem Reed jumped up and shouted, "One man don't prove nothin'." Then collapsed back in his chair and started hacking.

Okie Barlow shouted from the floor, "If it wasn't for Hi-ro-shim-a, we'd be eaten rice and drinking sake, and now we gotta deal with them damn Rus-kies."

Chee was struggling to help the Navajos make the connection between uranium, radioactivity, and lung cancer. There were hundreds uranium mines on the Colorado Plateau and many of them were on the Navajo reservation.

Russell Yellowhorse shouted, "*Dine*, what about the *Dine*?"

Chee stood and calmly asked Michael, "Yes, what about the Navajos?"

Michael turned to another tab in his binder, "Our best estimate is that over seven hundred Navajos worked in the uranium mines and mills in the Four Corners area, and since we started testing in 1945, fifty-four have died from lung disease, and another three or four hundred are incurably afflicted."

As Chee translated the Navajos fell silent, the arguing stopped.

Then Russell shouted, "What about Norman Benally?"

Michael looked helplessly at Chee, who nodded. He pulled an envelope from his pocket, "I have the autopsy done on Norman. He died of lung cancer from working in the mine."

It took Chee several minutes to calm down the Navajo miners. They were hard working, loyal employees who had been misled. They were shocked and angry, and struggled to hold it inside.

Myron Jones stood and waited patiently. As the Navajos settled down, he began, "Michael, as you know some of the folks in this room have worked in uranium mines. This is horrible news. Please understand we will work with you in any way we can to get the PHS to do their job. Now I don't feel right about moving off the uranium issue, but," Myron caught himself.

"I understand," Michael agreed. "And yes, I can discuss the hazards you face from the acid plant."

Every eye fell on him as he turned to another tab, "I will try to make this brief and only mention the five greatest hazards. Let's start with the ore crusher. Iron pyrite dust contains crystalline silica, which causes silicosis, a form of lung cancer and as you have learned from the recent fire, the dust is highly flammable. Second, the sulfur dioxide fumes that pour out the stacks and blanket the town are called acid rain. They are toxic and cause serious lung damage. Acid rain also corrodes metal, kill the trees, and blister your insides. Third, the vanadium pentoxide fumes from the converter are poisonous if ingested and cause bleeding from the nose, and serious lung damage. Fourth, finished concentrated sulfuric acid is also carcinogen. Inhalation of the fumes or mist can cause lung cancer, and it will peel the hide off an armadillo." He paused and smiled slightly as he went off script.

"Finally, as you are all aware, the drainage from the tailing ponds kills everything in the river and contaminates the water."

He closed the binder, cocked his head to one side, "The tragedy, the damn tragedy of it all," his voice broke, "is that most of this can be prevented with proper ventilation, maintenance and a few simple precautions."

Arguments broke out in small groups around the room. Susan Clever banged her gavel, trying to restore order.

Trish Vanderville strolled up to the podium, her hair snaked out like medusa, "It's time we stopped talking and did something about it," she fumed, "I teach most of your children at the school, and they are wonderful children, but it is a crime to raise them in this filthy town. As Roy and I say in our song: *Mother Earth is crying loud, Time to organize the crowd.*"

Part of the crowd started to sing along with Trish.

Okie Tom shot up and pointed at Trish, "I didn't know you allowed commies in Rico."

He pulled a spoon and a tin plate from somewhere and shuffled around the room like a dancing bear, banging the spoon on the plate, "That's the sound of an empty plate, get used to it, no Argentine, no jobs, no dinner!"

Lightening struck inside Johnny. In an instant he was a snarling alpha male, nose to nose with Okie Tom, "You want it now or later, asshole."

They circled, predator and prey, nose to tail. Off to the side, Roy slipped quietly from his chair and eased his way into a group standing behind the other Barlows.

Okie looked over at his brothers and gave them "the nod", a second too late. As the Barlow bothers sprung up, Roy put a hand on top of their heads and slammed them back in their seats.

Like a gambler reworking the odds to the last card dealt, Okie changed his plan. "Well now," he laughed fawningly, "You must be Johnny. We've been hearing a lot about you and Elvis there. I hope you understand, you are marked men, just like L.B. Birch."

Johnny rubbed his forehead, "the go-sign" he and Roy had used since high school. In a blur, he had Okie in a hammerlock and dragged him out the door. As he disappeared, Okie gasped, "No Argentine, no jobs, no Rico."

Roy followed, lugging the other Barlows out like cedar posts in a double headlock. They weren't gone long. Roy stooped back under the doorway and stood guard. Moments later Johnny returned, his left hand wrapped in a bandana. He looked around the room sheepishly, as though he had just woken up. Where in the hell did that come from? He wondered. Roy was shaking with excitement as he combed out his ducktail. An inscrutable smile crossed Polly's face. Most of the group were stomping and cheering.

Recognizing the opportunity, Brad jumped to the podium, "Johnny, you've been tight- lipped for quite awhile. If you got anything to say, this would be a good time to say it."

Johnny hesitated for the longest time, images of Norman fighting the bad wind, delivering his medical record to Michael, doing what was right, raced through his head. Then he stepped up, wearing a goofy grin, "Well, here goes. Michael, I agree with what you've said, but I think it cuts even deeper," he paused and looked around the room, trying to decide if he should continue.

Trish gently tapped a tambourine she'd brought for the occasion. Susan Clever gave up on controlling her.

"Last week, most of us helped save the plant from burning down," Johnny continued, "and I can tell you first hand it wouldn't have taken much to prevent that fire. The cost would have been peanuts compared to the rebuild." He looked over at Polly, who smiled broadly. "In the coal camp," Johnny continued, "where I come from, there were bad guys and they broke the law: somebody got beat-up, somebody got shot, and somebody got killed. And sometimes, if you were guilty, you went to jail. In Colorado, there are bad guys and they break the law, but it's different. It's harder to see things like: poor maintenance, poor ventilation, toxic fumes, and radioactive dust. Oh, the effects are just as violent and they have the same stench of greed about them, but there doesn't seem to be any laws against them. I am not sure how we change things around here, but, I am damn sure gonna try."

Like a tidal change the mood shifted. Even those who didn't plan to sign a petition were impressed with Johnny's comments.

Polly's eyes sparkled as he sat, "Well, where did that come from?"

"Don't know, but it feels right."

Trish Vanderville, tapped louder on the tambourine as she step up to the podium, "As I was saying, earlier, it's time to organize the crowd. The first thing we need to do, trust me on this, I've been in and out of more jails than Mother Jones, the first thing we need to do is to form a coalition, a coalition of local citizens that want clean water running in the Dolores River. And Roy darling, I give you permission to smack anyone that tries to interrupt me."

Roy smiled and flexed his biceps proudly.

Maggie jumped up, "I so move."

Polly was right behind her, "I second the motion, all in favor."

The ayes dominated. Those that opposed kept quiet.

Roy wandered around the room intimidating dissenters.

Then Susan Clever burst forward, "Trish, my dear, you are so out of order. This is not democratic at all."

"Listen sister," Trish hooted, "the last time you had a glimpse of democracy was when you were voted in as sergeant of arms in grade school, so sit down you Fascist."

"Trish, don't we need a name for the coalition?" Maggie asked.

"How about CCW, the Coalition for Clean Water."

Most of those in crowd applauded the name and started to chant, "CCW, CCW, CCW!"

"Alright folks, we've got a coalition and a name," Trish rolled on like the Columbia River, a born-again rebel and a political genius. "We are now an organization, Halleluiah! Halleluiah! Next, I recommend, we of the CCW sign a petition to protest the Argentine's pollution of the Dolores River and send it along with petitions from Dolores and

Cortez to the PHS and the Game & Fish Department. And finally, I recommend," banging her tambourine, "that the Mayors of Dolores, Rico and Dove Creek present these petitions to the next meeting of the Dolores County Commissioners. Let's roll!"

There were a few mumbling dissenters and some who silently opposed the recommendations, but no one with the courage or the smarts to take on Trish when she was rolling.

Bill Meyer, the mayor of Dolores spoke next, "We've got a petition going around Dolores, and we would be glad to join forces with you on either of your recommendations."

Myron Jones rose, "Speaking as the mayor of Rico, I agree with all of these recommendations, but I'd like a show of hands. How many of you agree?"

A straw vote was taken and a slight majority approved. The Navajos abstained.

Susan Clever, who had voted against all the recommendations, stood rigidly by the podium, glaring at Trish, "Do any of you remember L.B. Birch? He was the Superintendent of the Argentine before Chester Ratliff. L.B. was a decent, God-fearing, family man. You tell them Val, you tell them."

Val Brown jumped up but got so excited she choked on her words. She swallowed hard, cleared her throat and started over, "The Birches were only here a year or so. The *Dolores Star* said he resigned, don't you believe it. He started fussing about the way things were being run and, that's when Stanley T. Pritchard for you flatlanders, sent him packing. So those of you that work for the Argentine keep that in mind before you sign them petitions. Hell, Okie Tom's probably down in the Burley Building right now filling in Stanley."

"No, that ain't true," Roy blurted out, "me and Johnny sent him home to Dove Creek, and he ain't driving."

Again there was stomping of feet and cheering. A coalition had been formed and plans were agreed on.

As the meeting was ending, Chee stepped to the podium and searched for the right words, "We Navajo miners abstained earlier, but we just had a vote. We will sign the petition, tonight. We decided this before meeting with the chief, which is not done, so this may be good-bye. We must all fight the bad wind in our own way."

And so they headed home. The future was uncertain, yet the scale had tipped ever so slightly toward change. No one had been prepared for the fireworks that evening. Few fully understood what had happened. It needed to be raked and sifted, pulled apart and put back together. One thing was clear; Rico had shown grit. Some mining towns had gotten stuck in union-management violence; some had quietly accepted their colonial status, and some had just disappeared. Rico was a town struggling hard to become a community. Each faction had different needs, different values, and different levels of commitment. Like a tug of war, some would pull longer and harder than others.

The good folks of Dolores wanted clean drinking water, but risked little in the fight. For the sportsmen in Cortez, it was all about fishing. They wanted the river restocked but they too risked little in the fight. Chee like most Navajos worshipped nature and couldn't understand why white men wanted to pollute the rivers or harm their own people. They had put all their chips on the table. Michael, like David in the Bible, had flung a stone in the eye of the governmental Goliath and put his career on the line. He was special, a disciplined scientist with a passion for public health and safety. And then there were the mine workers like Johnny and Roy. They knew the pollution was senseless, and were tired of putting themselves and their families in danger.

And a few like the Fagans, Polly, and Hartley Lee were similar yet different. For them the wilderness was a sacred place that fully engaged them: salmon pink sunsets, the first snowfall, the symphony of sound and light in an alpine stream, the pungent resin of pine on a hot day, dancing Columbines in a light wind, a full moon rising, all of it connected, all of it part of a grand, timeless mystery. The wilderness, they felt, should be valued and protected, and yet their views held little sway. You couldn't quantify this or win a commercial argument with it. Were they born too early or too late? Somehow they were out of synch with the rest, wonderfully out of synch. And yet, truth be told, their values ran deeply through many of the others, muted, disconnected, unarticulated, yet part of them, at some deep, subterranean level.

Jim Rychtarik, now in his eighties, had spent most of his life prospecting and working claims in the high country. Spiritually and aesthetically he flew closer to the ground, a devout Catholic, but cantankerous as hell. With Jim, religion was one thing and Mother Nature another. He'd shoot claim jumpers on sight, just wing them you understand, but he lived in harmony with nature. He thrived off the challenge.

And so they headed home, a coalition with a plan, a measure of hope, and a budding sense of power.

Shiprock

Chapter 30

WIDOWS OF COVE

The wind blew sweet and clean across the flat topped ridge of the Lukachukai Mountains. Below, red rock buttes and mesas stretched across the high desert like crumbling fortresses from an ancient battle. Over fifty uranium mines were cut into the steep-sided canyons. Down on Mesa One, a dirty wind snaked out the shaft. Five miners had died from lung disease. Many of the older ones were not well. To the east, a red cloud tumbled down BIA Route 13.

The black hearse pulled over near the base of Shiprock, a forty-million-year-old volcanic pinnacle shaped like a Clipper Ship, rising eighteen hundred feet above the desert floor. Like the Grand Canyon or Yosemite Valley, Shiprock is imposing, ethereal and astonishingly beautiful. The Navajos call it *Tse Bit'a I*, or "Rock with Wings," after a legendary bird that brought the tribe from the north.

The procession of cars and pickups came to a stop behind the hearse. Sagebrush and juniper covered the baked flatlands. Tumbleweeds piled high along a barbed wire fence that ran along the red clay road. They left Ertel's an hour earlier, headed for the Native American Church in Red Rock and then on to the family plot.

Chee gazed at Shiprock and smiled. A gentle wind blew at his back.

Johnny watched, "You seem happy."

"Oh, I am happy and I am sad. I will miss my Father, but the spirits are with us. This morning in Ertel's Chapel we saw a clipper ship crossing the bar beneath a marbled blue sky. Now, here at Shiprock, we see a clipper ship, sailing beneath a marble blue sky. These are not accidents. The stars are aligned; the wind is at our back. That is *nilch'I*, the Holy Wind. These are good signs."

Johnny rubbed his chin and wondered how he missed so much of Chee's spiritual world, "I will say this, if God lives in nature, Shiprock has to be one of his homes."

Chee smiled to himself. He struggled with the finality of death. As a boy, he was told there was no heaven, no hell, nothing after life. He believed this until he was assimilated into Western ways and the teaching of the Native American Church. He liked the idea of Norman crossing the bar. He wanted his father to be somewhere, on a clipper ship or on the back of a galloping stallion. He wanted his father back. He stole a final glance at Shiprock, then pulled ahead of the hearse, taking the lead as the road disappeared into the dust cloud they'd kicked up, deep in the Navajo Reservation.

Johnny sat in silence across from Chee, slowly drawing a file across the blade of a hatchet. He'd fitted a new handle on it last night. As he switched hands he took a sip from the pint he been working on all morning.

"You want to talk about Norman?"

Chee smiled for the first time, "He's ready for the big ride. He grew up on horses, long before pickups."

"Norman had a gentle hand with the horses."

"He talked to 'um as well."

Johnny switched hands, struggling to stay calm, "I saw some of that."

Even during the day Johnny was haunted by images of the bad wind. It was all running together, reality and illusion.

As they swung around a sandstone bluff, Chee pulled over to let the rest of the procession catch up. "See that triangular shaped point just below the skyline to the north?"

"The sloping, red rock point?"

"That's it. That's Mesa One, a Kerr-McGee's mine."

"My dad and I worked there years ago. We'd drill holes, pack them with dynamite, and let it rip, then race back in and get to mucking, yellow dust everywhere, no ventilation. We'd fill up the ore cars, then they'd haul them out and dump them down a shoot. The tribal council says there are over a thousand mines on the res, most of them small, unventilated shafts, called dog holes."

Navajo mine workers at Kerr-McGee Mine, Cove, AZ

"Did you do any bracing or roof bolting after a blast?"

"Are you shitting?"

"There was never any of that. Sandstone is soft. Lots of hanging slabs. Lots of injuries from cave-ins and rock fall. You won't see many old miners on the rez that aren't hobbling around with some kind of injury."

"Sounds like West Virginia."

"You and Roy were good last night. How's your hands?"

Johnny stopped filing, and open and closed his bandaged hand, "My ring finger started to swell after the meeting, and Roy had to cut off my ring with a hacksaw. Damn near took the finger as well."

"They say you knocked Okie Tom out."

"I am not sure if he was out, but he did stay down. I thought you didn't approve of fighting."

"Oh, I approve, but it's not a hobby for me, like it is for you and Roy. I will fight injustice, one way or another.

"Jesus, you sound like a missionary."

"Michael put his ass on the line last night."

"Yes he did."

"So did you, Crazy Bull."

As they bounced from pothole to pothole, a stream of images ran through Chee's mind: a young boy on a pony racing his father up Red Mesa, building a campfire at sunset, watching the stars come out, laughing in each other's arms. He'd never known his mother. She must have been a wonderful woman. She chose well.

His father was a traditional Navajo in some ways, but you never knew. He had more Gods and demons than any one tribe, and more irreverent moments, and more damn fun than any Indian on the planet. His wife was a Lakota Sioux. They met at an inter-tribal gathering in Wyoming. Her spiritual world added a rich overlay to the Navajo beliefs of harmony and goodness, of *hozoho*.

"You going to Shiprock?"

"First, I've got a meeting with the Cove Chapter elders, to see if they'll go along with the petition."

"If they don't?"

"We Navajos understand politics, *bilagaana*. We'll find a way."

Chee smiled as he caught site of the limo in the rear view mirror, "Yeah, Norman will definitely like the limo. And he'll like the strangeness of a Navajo "Road Man" and a *bilagaana* funeral. That will really crack him up."

The funeral ceremony was held in the Native American Church (N.A.C.) in Red Rock, and involved both Christian and N.A.C. prayers, songs and eulogies. From there, the procession headed for the family burial plot several miles away. In about ten minutes they ran out of road. A cluster of wood frame houses and stacked log *hooghans* spread out below a massive red rock alcove. Dozing dogs stirred beneath a shade tree in the yard, circled around and began barking. This was the Benally homestead. In the backyard, beyond a horse corral and grazing sheep, a footpath disappeared into the brush. A corn patch stood on the sandy bank of a wash. In the distance, a fantasia of red sandstone buttes, window rocks, and comic hoodoos towered above the parched landscape. The hearse was caked in chili-red clay.

Chee calmed the dogs and ambled over to Keenan, "The burial ground is just up the path, a quarter of a mile or so. Take her slow."

Keenan smiled kindly. He had arranged many funerals on the reservation and nothing surprised him anymore, "We'll drive as far as we can and carry the casket from there."

Chee bushwhacked, as the hearse lumbered through the underbrush and in and out of a several dry washes. Those in the procession followed on foot. There was a loyal group from Rico: the Fagans, the Yellowmans, the Lees, Myron Jones, along with Jim Rychtarik, Lucy Fahrion, and Roy and Trish. Michael Burns and Doctor Chavez had also come along. The last hundred yards, the hearse bogged down in the sand, the pallbearers were called on.

The burial ground was a small clearing skirted by piñon pines, a hidden bench hanging off the Lukachukai Mountains. The air was spiked with sage and the pungent smell of juniper. Elevated burial mounds marked with small white crosses and dotted with clusters of artificial flowers spread across the burial grounds. Some graves had knee high picket fences around them. Some had headstones carved out of sandstone slabs. Children's graves were often littered with their toys. Sagebrush rolled along between the graves.

Lowering straps hung across Norman's deeply cut grave, nearby a hobbled horse grazed quietly. A gathering of family waited patiently on one side, mostly Norman's brothers and sisters, their children, and their children's children. A group of older women huddled next to the family, heads down, praying silently. The Rico group assembled on the other side. The pallbearers set the casket beside the grave and stepped back.

The Road Man was gaunt and stooped at the waist. A red headband held his hair to the side. His face was cut from stone. High cheekbones and a prominent nose set off deep sad eyes. A silver crescent hung from a squash necklace, along with bunches of turquoise beads. He built a small fire of cedar branches near the grave and fanned the smoke to bless the grave. Then he danced around the casket, chanting and flourishing eagle feather fans. A drummer joined him in the ceremony. He opened the casket and tucked a beautifully woven saddle blanket around Norman's feet.

Chee smiled down on his father. This was a joyful day; the coughing and pain were gone. Norman stared up into the sky with a proud, fearless smile. A tattered old saddle lay at the foot of the casket, his right hand held a bridle, ready for the last ride. The others began to file by, saying their goodbyes, some left an offering.

Johnny's head was spinning, the visions of the bad wind had returned, fueled by a pint of cheap whiskey. He steadied himself, then drifted, then was nudged awake by Chee. It was time for his offering. He laid the hatchet in Norman's left hand and whispered, "Keep the fires burning."

The Road Man flung water on the casket as it was lowered. When the ceremony was over, Norman's closest friends stayed to finish the burial. Each shoveled dirt on the casket, shared a story of Norman's life, his courage, his humor, and passed the shovel along.

Later the mourners gathered in the auditorium of the Cove Day School, which served as the theater, the basketball court, the lunch room, and for any other gathering in Cove requiring a large room. They waited patiently in line to express their sympathies to Chee, often sharing a fond memory. Then they got into the food line. The serving tables were set up below the basketball net in Norman's honor. He had been a star player as a young man. They were heaped with traditional Navajo food: blue corn meal pudding, kneel down bread, fried, baked, and barbequed mutton, yellow squash, Navajo tacos, watermelon, and lots of cake and ice cream.

The group of older women was the last to offer their condolences. They wanted to speak with Chee before he left. The reception lasted for three hours. Chee thanked everyone as they left. Towards the end, he joined the older women in a corner of the auditorium.

"Hello Chee, I am Dorothy Russell. I will speak for my friends. My husband worked for U.S. Vanadium here in the valley. We all knew Norman as a young man."

"Thank you all for coming."

Dorothy introduced the others. They were beautiful women, weathered by time; their silver hair pulled back, their sad faces soft as deerskin. Turquoise petite point necklaces hung over black velvet blouses.

Chee looked deeply into Dorothy's eyes and hesitated, "Are you alone?"

"George, my husband, died last year. The others are not well. We have many questions. Perhaps, this is not the time?"

"Please, speak freely."

Dorothy huddled with her friends and they conversed in Navajo, then she turned to Chee.

"How did Norman die?"

The Rico contingent was starting to leave. Roy and Trish waved goodbye from the door, not wanting to intrude.

Chee spoke to Dorothy in English, not sure how much she wanted to share with the others. "It began several years ago with gasping and shortness of breath, then he started coughing up blood, then the headaches and great pain on the left side, and then he was gone."

He stopped to let Dorothy translate in her own way. Michael Burns and Doctor Chavez were preparing to leave. He excused himself and met them at the door, "Thank you for coming and thank you for everything you have done. If you have a moment, I may need your help."

As they walked over, Chee filled them in. As Dorothy explained Norman's symptoms, the women moaned and shook their heads. When she finished, Chee introduced Michael and Doctor Chavez.

Dorothy brushed away her tears, "My husband, George, died like Norman, and their husbands are suffering in the same way."

Chee turned to Doctor Chavez, "Can you help them?"

The doctor sat next to Dorothy, "If you want my help, I must ask two questions."

Dorothy explained to the others. They agreed to continue.

"First, did they smoke, and second, how long did they work in the mine?"

The women talked it over carefully for several minutes, Dorothy summarized.

"All started in the mines in the late forties. One smoked as a boy, the rest did not. Most are trying to drink through the pain," she said sadly.

"Sorry, for asking, but most, probably seventy-five percent, of lung cancer comes from smoking. The rest we believe, in this situation, comes from mining exposure. The autopsy showed that Norman died of oat cell carcinoma, a cancer that had metastasized; excuse me, that had spread from his lungs to his brain."

Again the women put their heads together and discussed his reply.

"What is this word, exposure?" Dorothy asked.

The doctor turned to Michael, who continued on, "Uranium is radioactive, it gives off a gas. You can't see it, you can't smell it, but it is evil. This, we believe, killed Norman and others."

Dorothy turned to Chee, "Radioactive?"

"*Bideezla'na'alkidgo.*"

"*Bideezla'na'alkidgo,*" she said slowly, turning to her friends. They all whispered it.

She locked her hands together. Her head shook from side to side. "If this is true, a bad wind blows from the mine. Why weren't we told? Ten years ago the PHS came to Cove, to this very school, and did their examinations and their X-rays. Why weren't we told? Every morning our husbands climb up to the mine. We hear the thunder of the jackhammer, the roar of the blast. We see the black wind fly from the mine. At night they come home covered in yellow dirt, in *leetso*. Why weren't we told?"

The anger swept over the woman like a rash. They chanted a song of exorcism from the Wind Way Ceremony, hoping to drive off the bad wind.

Chee looked helplessly at Michael, "Why weren't we told?"

Michael was again at a crossroad, the second day in a row. He was not turning back.

"Ladies, you will not like what I have to say, and neither do I. But please hear me out."

The chanting stopped; all eyes were riveted on Michael.

"I work for the Public Health Service and I have been involved in many of these studies."

The women were shocked. Dorothy started to interrupt.

Chee stopped her. "Hear him out, he can help us."

Michael stared down at the floor, unable to look them in the eye, "At first, we didn't know what caused the illness. We were told to study it. Now we can say, since 1945, sixty-five miners have died from exposure to uranium on the Colorado Plateau; forty-one since 1960. We were told to continue our studies, but to say nothing. The Atomic Energy Commission continued to build their bombs." He raised his head, tears ran down his face, "This was wrong. Last night, I shared this information for the first time in Rico, and I am working for change. I am truly sorry."

Dorothy was touched, but persisted, "How many examinations have you done?"

Michael grimaced, "Over thirty-two hundred."

Shock and anger washed across the widow's faces. No one spoke for several minutes.

Finally Chee broke the silence, "Michael showed much courage in speaking out. Now we know. Now, we must warn the Navajo people. I will join you."

Chee suddenly realized he was changed. There was no turning back. This was his fight and he was ready for it, but it would change the course of his life.

Dorothy searched the faces of her friends and asked them in Navajo, "Are we willing to fight the bad wind that comes from the mines, the *bideezla'na'alkidgo* that kills our husbands?"

They talked among themselves for several minutes. Then turned to Dorothy and nodded agreement.

Dorothy put her arm around Chee, "And Big Thunder, we, the widows of Cove, will help you. And Michael we forgive you. I do have one last question, if I may."

Michael nodded solemnly.

"Why have the deaths increased?"

"There is a latency period of approximately fifteen years once exposed."

Dorothy looked questioningly at Chee.

"He means it takes fifteen years before..."

The woman embraced and returned to a chant of exorcism:

> *Where my kindred dwell, There I wander.*
> *The Red Rock House, There I wander.*
> *Where dark k'eet'aan are at the door, There I wander.*
> *With the pollen of dawn upon my trail, There I wander.*

At the Cove Day School that afternoon, the chrysalis of Chee's *bilagaana* assimilation, an assimilation that had held him earthbound, finally cracked open. As a boy he left the reservation, left the simple nomadic life of the shepherd to become an industrial worker. He had tried to assimilate, to adapt but he was caught between two worlds, bound up, encased in a chrysalis of his own making. It was time to reconnect with his roots, too embrace his native culture, a culture of *hozho* and *Sa'ah Naaghai Bik'eh Hozhoon**, a culture that teaches humans how to live with each other, how to live with the animals, the plants, the earth, and the universe.

He stretched his monarch wings and flew high over Shiprock, then sailed back over the Lukachukai mines. His metamorphosis was sudden and unexpected. He was no longer the lead jackhammer in the mine. He was a Navajo warrior fighting for the *Dine' bikeyah*, fighting the "*bideezla'na'alkidgo*". He would cure the land of the bad wind. He would keep the fire burning on the mountain. He would do what was right. He would live in balance and harmony.

Rico from Beamis Flat up Silver Creek, Blackhawk Mountain, right

Chapter 31

MAKING THE WIND BLOW

Spring 1964

"Politicians are like weather vanes; our job is to make the wind blow." —David Brower

Nature does its best work in the spring, a time of renewal, of budding and blooming. When the sky is blue and wildflowers paint the meadows, anything is possible. And so it was in Rico, as the Clean Water Coalition (CWC) fanned out across the Western Slope, to harness the drafts and breezes of dissent, to create a windstorm of opposition against the man. The campaign was launched on every street corner and crossroad: picnics and pool rooms, fairs and festivals, ballparks, and grange halls. It was spring and they were off, preaching their gospel of hope and renewal.

Most people opposed air pollution, acid rain, and radioactive dust. To the astonishment of the CWC, there were no laws against air pollution in those days. They focused, therefore, on water pollution, carefully mapping out potential allies: the activists and fence-sitters in Rico, the downstreamers that relied on clean drinking water, the recreationalists that came to fish and play, the Navajos that came to work, and the various agencies responsible for land management and public safety. Each faction valued the wilderness in a different way, each had a stake in cleaning up the river, and each was prepared to take a certain risk in solving the problem. None were rich and powerful, but collectively they could make the wind blow. The plan was founded on this narrow ground, and the CWC marched forward, packing town halls, passing petitions, filing injunctions—working the system.

Polly and Johnny petitioned folks as they came into Lucy's Lounge. Roy shadowboxed in the corner doing his best to intimidate. Maggie and Trish prowled the streets of Rico like honey badgers. Most of the support for the cleanup, however, came from downstream communities. The Fagans met with the Cortez Chamber of Commerce and gained their agreement to join forces with Rico. Myron Jones met with Bill Meyer in Dolores and they too agreed to join forces, and set up a meeting with the Game and Fish Department. Chee met with the Tribal Council in Shiprock. Hartley and Irene Lee met with the Dolores County Commissioners and obtained their agreement to support the clean-up, while Hartley continued to rail against pollution in the *Dolores Star*.

The Game and Fish Department and the Public Health Service had done water sampling along the river, but neither released their findings. Michael Burns pushed the PHS to state the minimum water standard. There were internal debates between both departments as to what the exact standards should be and how to enforce them. Some of this was good science; most of it was bad politics.

And so, the CWC pressed on, sustained by the rightness of their cause, yet buffeted by the winds of resistance. The Argentine grudgingly agreed to dry stack the tailings, rather than dump them into the tailing ponds. No one really understood this, but it sounded positive. Various petitions had reached the Denver office of the Game and Fish Department and they had instructed their field personnel to maintain constant liaison with the Cortez Chamber of Commerce. This too seemed positive. And Bill Meyer, the Mayor of Dolores, with petitions in hand from Rico, Dolores, and Cortez, personally appealed to the PHS. They reluctantly agreed to continue studying the situation. Short of taking action, this was their crowning achievement.

The first flowers bloomed in April. The Game and Fish Department refused to restock the river until more tests were conducted. Their latest sampling had shown the mayflies and stoneflies, basics of the trout's diet, had been killed from the acid run off and that no fish had been found. Soon after, the Dolores County Commissioners declared that the Argentine must pay seventy thousand dollars in back taxes. The CWC was clearly on the move, yet the acid plant ran around the clock and neither the air nor the water was clean for long.

Anyone who spent the winter in Rico knew that spring was a wonderful time, but summer, summer was what you lived for. That summer, pretty Polly Carnifax gave birth to Julie, a ten-pound darling. And once the snow was off the high mountain trails, the Rychtarik and Snyder girls saddled up and galloped off. Rico girls were a hardy lot. They loved to dress up, they loved a party, but they also enjoyed chopping wood, exploring old mines, and trail riding. From the Highline Trail, the girls looked out onto Utah, New Mexico and Arizona, a grand vista. When they got back to the ballpark, they barrel raced until dark.

A few more bullet holes were added to the ceiling of Lucy's Lounge at 1:01 am on the morning of July 4th. Nothing serious, just a few local patriots bringing in Independence Day. The Fourth was the biggest day of the year in Rico. The sun was shining and the sky was a blue. American flags fluttered from every storefront and bunting was draped across every railing. Rico was dressed for a party.

The Women's Club posse scurried about, checking to see that the parade, the cake sale, and the Old-Timer's Picnic were ready to go. They had carefully prepared for the day, but understood that events would quickly take their own course. The parade was forming-up on Soda Street. A happy pandemonium prevailed. The Four Corners Community Band tuned up on the back of a flat bed truck, while young boys with firecrackers were on the prowl, spooking horses and waking old-timers snoozing in lawn chairs. Cowboys with a Circle K Ranch banner tried to steady their mounts. Teenagers from the Book Club put the final touches on the Rico Library float, a replica of the Galloping Goose. The original Goose was a car engine and body converted to rail, a humorous but beloved contraption. It replaced the steam engine in 1951 when the Rio Grande and Southern went broke, carrying the mail and a few passengers. They say it waddled and honked like a goose. Two live geese in pens were tied to the front of the float and honked on cue.

A four man Color Guard from Cortez in full dress uniforms scurried around setting up their sound equipment on a Willys Jeep. They would lead the parade and present the colors at the Parade Marshall's stand in front of the Burley Building. The Marshall this year was Mr. Stanley Pritchard, the President of the Rico Argentine Mining Company. Not a popular choice.

The Veterans of Foreign Wars assembled behind the Color Guard. One surviving World War I soldier would lead the VFW. Myron Jones, Charlie Yellowman, Jim Starks and Becky Riva made up the World War II veterans. Myron, a Second Class Torpedo man, wore a standard Navy "crackerjack" jumper, bell bottom trousers and his sailor's cap at a jaunty angle. He'd been on a PT boat in the Pacific firing on Japanese destroyers. Sergeant Charlie Yellowman, a Code Talker on a B25, wore a colorful uniform, a red garrison hat and a gold-colored shirt. A silver squash necklace hung from his neck. Code Talkers sent and received confidential radio transmissions during the war. During the battle of Iwo Jima, Navajo code talkers sent and received eight hundred messages in two days. Major Howard Connor, 5th Marine Divisional signal officer said, "Were it not for the Navajos, the Marines would never have taken Iwo Jima." The same has been said about many of the battles in the Pacific.

Jim Starks, Second Class Machinist Mate on the carrier Midway, wore the same uniform as Myron. Becky Riva, a Warrant Officer in the Women's Army Corps, the WACs, served in New Guinea. She wore an olive drab skirt and jacket with a hobby hat. There were five veterans from the Korean War. One of them was Johnny Carnifax, an Airman Second Class. He wore a dress blue jacket with brass buttons. All the veterans were highly decorated.

American pioneers that settled the West in the nineteenth and twentieth century crossed many frontiers, each with its own hardship and challenge—winter storms, desert heat, starvation and death. The frontier shaped their character, like a blacksmith forged and hardened iron. This rarely occurred in the relative comfort of the east. Some pioneers never got beyond St. Louis, some turned back along the way, and many died on the trail. Those that kept moving west were fearless—scrappy, self-reliant, independent. Like the frontier, war also shaped people's character in ways that civilian life never could. Just as living in the high country of Colorado hardened their character in ways that didn't happen in the flatlands. The veterans in today's parade and their ancestors weren't saints, but they were a special breed, leaders wrought by the frontiers they'd crossed, the wars they fought, and the high country they endured.

Strung out behind the VFW were the Masons, a Navajo Dance group from Shiprock Day School, the Dolores High School Marching Band and dozens of kids on bikes and trikes decorated with flags and crepe paper streamers. Jim Rytarick, the last of the true mountain men, wore fringed buckskins and a coonskin cap.

As he waited for the parade to start, Myron pulled Johnny aside, "For a flyboy you look sharp there, Johnny."

"I feel a little foolish, but Polly insisted I join in. Haven't had it on since 1953. For a swabby, you don't look so bad yourself there Myron."

"I haven't seen your buddy Okie Tom Barlow around town lately; you must have whacked him good."

"Well, I caught him with a left hook," Johnny chuckled. "They say his jaw was wired shut and he's was eating through a straw for awhile. Ain't that a shame?"

"How your hand?"

"The knuckles fine, but my ring finger really swelled up."

"You don't say."

"Roy had to cut off my wedding ring with a hacksaw, damn near took off the finger, just a small scar now."

"Say, Johnny, if you don't mind, I'd like to talk with you after the parade. Maybe down at the picnic? I am not seeing much progress at the acid plant."

"There hasn't been. I'll find you at the picnic."

At the ballpark south of town, a crew from the Rico Fire Department was hard at it, cooking shoulders of prime Stoner beef for the picnic. Like every high mountain town, Rico had its traditions, and nothing was more sacred than the preparation of the meat for the Fourth Picnic, an offering to the Gods and to hungry miners. Telluride had its barbecued beef, and Dunton had its venison or whatever, but only Rico baked their beef, Hawaiian-style underground. The firemen worked through the night, stopping only for the occasional beer. Shoulders of beef were seasoned with a secret potion, wrapped in burlap, and packed into a large, stainless steel oven that was lowered into a deep pit. Coals from a bonfire were raked into the pit, and covered with dirt. Every four hours or every time they finished a case of beer, whichever came first, they'd uncovered the oven and checked the meat, adding sausage, and potatoes as they went along. The spicy aroma drifted up Glasgow Avenue reminding everyone that the picnic followed the parade.

Independence Day was a festive day. Commerce stopped and the national birthday party began. On the Fourth, the town of Rico became a community, and America, a diverse collection of cultures, tribes and nationalities, joined hands as a united nation.

At eleven o'clock, the fire siren rattled the windows throughout the town, announcing the start of the parade. The Color Guard led the parade, moving down Soda Street onto Glasgow Avenue and marching briskly south. The VFW got the greatest applause, even greater than the volunteers on the fire truck throwing candy to the kids.

As they reached the Burley Building, the four soldiers in the Color Guard halted. "Right face!" barked the leader. They turned crisply to the right and came to attention again. "Present arms." The two soldiers on the outside brought their rifles off their shoulders and held them vertically in front of them. The two flag bearers in the inside held the colors. The American flag on the left was held straight up. The Colorado State flag on the right was dipped. The driver of the Willys Jeep behind them turned on the recording of the National Anthem. Everyone on Glasgow Avenue stood at attention and saluted the American flag, which fluttered freely in the breeze as the anthem rang out. No one blinked. Even the horses stood still.

> O say, does that star-spangled banner yet wave
> O'er the land of the free and the home of the brave.

Myron's jaw tightened as the VFW group stood at attention behind the Color Guard. He was a serious student of American history, having read everything in print on the Revolutionary War, The Constitution, The Bill of Rights and the biographies of every president. Abraham Lincoln and Teddy Roosevelt were his heroes. He was the last truly independent miner in town, owning and operating several mines in West Rico, the Yellow Jacket Mine up on Silver Creek and he had a partnership in the Silver Bell Mine in Ophir. He worked for himself, hired who he pleased, paid his taxes and played a major role in the community.

Today was a day to celebrate the American ideals of freedom and independence, yet Myron was troubled. Things had changed lately with faceless corporations and holding companies running most of the mines. Dependence had replaced independence and industrial workers had replaced the pick and shovel miners. Democracy didn't work so well anymore, and Myron still hadn't figured out why.

As the National Anthem finished, Stanley T. Pritchard, the Parade Marshall, stepped to the podium, "It is a great personal honor to serve as your Marshall. Today, is indeed a grand day for America, we are without question the greatest country in the world. Yet we are currently engaged in an arms race that will determine the future of the world. Make no mistake about it; the security of this great nation is at risk. But look to your left, then to your right, we, we the great citizens of Rico are playing a critical role in this battle. We are the foot soldiers of the Cold War." He droned on for another ten minutes and finally concluded, "So, let me be clear, freedom and independence cannot survive without American industrial superiority. Democracy can only thrive in an open market of unfettered capitalism. Keep up the good work and God bless you all."

A temperate applause followed. Myron caught Stanley's eye as the parade moved on. They both scowled. By now, Myron was fuming. In the past, he'd heard words like freedom and democracy mixed in with words like industrial power and unfettered capitalism, just another Fourth of July speech full of platitudes and bombast. Yet today it sounded evil. The acid plant was polluting the town and uranium mines were killing mine workers all across the Colorado Plateau. Most of what they had tried, to correct the problem had failed. The guarantees of the Bill of Rights, of due process and of the redress of grievances seemed meaningless anymore.

The parade ended at the firehouse where the auctioneer immediately started his bidding chant at the bake sale. The Four Corners Band assembled in the driveway and set off playing *Yankee Doodle Dandy*. The firemen sold beer and cokes and the Yellowmans were doing a booming business at their Navajo Taco booth. Most people milled about, chatted with old friends and shared fond memories.

Some wandered over to the Rico Hotel across the street, certainly one of the finest hotels in Colorado, though some would say it was haunted by old miners. Denver had The Brown Palace, a treasure of Italian renaissance design; Leadville had the Delaware Hotel, a Victorian adventure; Durango had the Strater Hotel with its grand Victorian gimcrackery; and of course Telluride had The New Sheridan Hotel, a sumptuous brick fortress where William Jennings Bryan, the Democratic Presidential candidate in 1896, gave his famous Cross of Gold speech. But on the Fourth of July in Rico, there wasn't a finer place to enjoy a cold beer or a fine French wine than the Rico Hotel.

The picnic at the ballpark would begin in an hour followed by a horse race and a softball game. The fireworks started at eight and the dance at nine. Myron bumped into Johnny in the picnic line and agreed to meet him at the beaver pond by the highway. Every imaginable salad, casserole, and pie were spread across the serving table at the pot luck picnic. Myron and Johnny loaded their plates and slipped away. They sat on a fallen tree beside the pond and ate quietly. The pond was crystal clear and teeming with life, cut off from the main flow of the river and catching several freshwater creeks that flowed directly into it. The beaver dam and the lodge were models of efficient engineering, just enough branches and mud to do the job, no more, no less. This seemed a good use of nature.

When finished, Myron wiped his mustache with his handkerchief, "Johnny, I am proud of the way you spoke up at the meeting in February. It was a long time coming, but you had your facts together. The Argentine never did much to prevent fires and your report sure pointed that out."

Johnny enjoyed the last bit of apple pie, "You know a good apple pie beats the hell out of small town politics. Thank you for your kind words, but what the hell, it's been five months since that meeting and none of my recommendations have been acted on. The plant is still a fire trap. According to Maggie, the insurance company is going to pay off their claim. Now ain't that some shit."

"Did you hear that Michael Burns was reprimanded and quietly reassigned to the Utah office of the Public Health Service? They heard he'd spoken out in Cove and at our meeting. They said it was a breach of national security. So, don't expect any help from PHS in the future."

As they looked up, they were surprised to see Chee leaning against a nearby tree with his thumbs anchored firmly in the front pockets of his Wrangler Jeans.

"Well, hello there, Thunder," Johnny roared.

There was a flurry of pumping hands and bear hugs. They were delighted to see him. Chee had moved back to Cove after his father died.

His hair was up in a traditional Navajo bun with a bandana around his forehead. He wore the big buckle from his rodeo days and silver bracelets on each wrist. A small American flag stuck out of his shirt pocket.

Chee flashed a wry smile, "It was good to see a Code-Talker in the parade, you know, here in the *land of the free*, and the long walk, *and the home of the brave*, that's a Navajo brave, you understand."

Acknowledging Chee's biting irony, Myron nodded, "Guilty as charge."

Before long, they were all down on their haunches fiddling with twigs and catching up.

"Are you making any progress down in Cove?" Johnny asked.

"I go from chapter house to chapter house, from Cove to Red Valley, to Shiprock, along with the Widows of Cove warning everyone of the dangers of radioactivity, of *bideezla'na'alkidgo*. We're going to meet with the elders in Window Rock next week to demand compensation for the families of those that died or are dying. We are also trying to get some congressmen involved, work the system as you *bilagaanas* say."

"Do you still have bad dreams?" Johnny asked.

"No, I have good dreams. I am working to drive off the bad winds and cure the land. This is good. I have more harmony and goodness in my life, more *hozoho*."

Johnny got up and leaned against a tree, "We are mighty proud of what you're doing. I can't say I've got a lot of *hozoho* in my life. I'm still having them bad dreams. You know the ones, racing like hell to keep the fire in the meadow and the fire on the mountain burning. Norman can't be happy with me."

"Is the CWC making any progress with the petitions and the meetings with the Game and Fish or the Public Health Service?" Chee asked as he drew a circle in the ground.

Still on his haunches, Myron whittled in silence, taking it all in.

Johnny shrugged, "Not much, Michael Burns has been sent away for speaking up, the Argentine is still running a dirty operation, and the river is dead. The Game and Fish Department keep taking samples, and they don't stock it anymore, but there have been no fines or arrests."

They both looked at Myron, "So, what do you think?"

Myron rubbed his chin and slowly cleared the ground in front of him. They waited for the longest time. They hoped he would give them an answer. Sometimes he didn't. He had to run the question through his mind several times and if he had an answer or the glimmer of an answer he might share it. By now he had several sticks, like tent pegs, in his hand he sharpened. "Chee, you Navajos are good with riddles, so you may be able to help me with this. I haven't got this worked out yet, but I'll give it a try. Things aren't right up here nor down on the Colorado Plateau. So, why is that? I'd say the first factor is the opportunity for profit, gold, silver, acid, uranium." He drove a stake in the ground, "None of us would be here without that. So, that's an important factor, the opportunity for profit, and depending how you work it, the opportunity for obscene profit. And if you're not careful, that draws blow flies, maggots, and lawyers. Are you with me on that?"

Johnny and Chee nodded their eyes fixed on the stake.

"And then you got these evil, all powerful, absentee mine owner that don't give a shit about the community," Myron drove another stake in the ground. "Sorry fellows, there is no excuse for that kind of language, but this one gets me real hot."

Chee nodded, "It wasn't like this before the *bilagaanas* arrived."

"No it wasn't, nor was it like this when my father, Lewis Jones, started mining up here. In the twenties and thirties there were over a hundred mining companies working from one man up to as high as two hundred and fifty men in a mine. You did your prospecting, you staked your claim, and you worked it. And if you got lucky, you struck it rich, pissed it away, and started over. I mean look at this pond here, ole Mister Beaver and his clan. They've staked their claim and they're working it. They take some and they leave some and they're good for the environment."

"Yes sir, Mister Beaver is a good fellow. Now I don't want to argue with you Myron, cause you're about the only fellow in town that could whip my ass, other than Chee, but them two stakes in the ground are pretty much what you'd find in the coalfields of West

Virginia, except we got another couple of stakes, the union and the Baldwin-Felts detectives, who played both sides against the middle."

Myron stepped back like a scientist in a laboratory, struggling to find a formula that could unlock a solution, "It's still a work in progress, boys, but I believe I am on the right track. Anyway you figure it; we're talking about greed—infectious, corrosive, malignant greed."

"Amen to that!" Johnny said cordially, though he was more disturbed than he let on.

They headed back to the picnic, three patriots with a cause. All went well the rest of the day. The Old-Timers Picnic was a great success, Rico beat Dolores in the softball game and the fireworks were the best ever. Disappointed that he hadn't won a prize in the parade, Tyke Swank got loaded. With American flags streaming from the handlebars, he rode his Harley into the dancehall, demanding a recount. Myron Jones calmly lifted him off his bike with one hand and carried him outside. He told Tyke he was welcome, but his bike wasn't. The Women's Club made three hundred dollars at the bake sale, and every Navajo cheered as they watched the movie, *The Battle of the Little Big Horn* at the company theater. The Indians finally won one!

In early August, the CWC finally got some good news. The District Attorney's Office in Cortez was getting involved. The following public notice was printed in all the newspapers.

"As a result of numerous complaints received from area residents along the Dolores River, State Game and Fish Officials, sportsmen and other interested parties concerning the critical pollution problems existing for more than twenty miles below Rico, Colorado, the District Attorney's office has contacted Mr. Stanley T. Prichard, president of the Rico Argentine Mining Company, in an effort to resolve this serious problem without the necessity of resorting to the courts.

The meeting is scheduled at Cortez, Colorado on August 9, 1964.

> *Respectfully,*
> *Mr. Wade A. Dillon*
> *District Attorney"*

Bubbling with insider information, Maggie met Polly and Julie for lunch the next day. She said the Emperor penguins had been in town, while extending her stomach and waddling across the floor to mock them. The penguins, the Argentine's Board of Directors, had come to discuss Union Carbide's plan to build an acid plant in Uravan, a plant that could put them out of business. They also discussed Chester's impending meeting with the District Attorney in Cortez. Maggie put Julie on her knee as she continued. She also reported that by some legal sleight of hand, the insurance claim for the ore house fire had been approved in spite of Brad's fire report and Johnny's analysis. And so it went.

Jack Stark, from the Attorney General's Office in Denver, flew in to represent the Colorado Game and Fish Department at the Cortez meeting. When questioned about the problem, Chester played the artful dodger to perfection. He said the plant would be shut down for the next two weeks to install a tailing disposal line. This, as he had said repeatedly in the past, would enable them to dry stack the tailings and minimize dumping them into the ponds. They also planned to line the ponds nearest the river, yet never admitting any responsibility for the pollution. Jack Stark concluded the meeting with the following statement, "We all hope for the best, but in the event that the proposed modifications and changes in the operation at the acid plant does not stop the pollution of the Dolores River, The Colorado Game and Fish Department, in cooperation with Wade Dillon, Assistant District Attorney, plans to file suit in District Court to stop the pollution and recover damages for fisheries' losses. This legal action was approved by Governor John A. Love and the Colorado Game and Fish Department."

Conspicuous by their absence were: Mr. Stanley T. Pritchard, and any officer from the Public Health Service. No one dared celebrate, but this was damn good news.

Throughout the summer the CWC and its angry members from Rico to Cortez complained about the river pollution to the District Attorney and the Game and Fish Department. Some felt it was more than leaching or the overflow of the tailing ponds. They felt the Argentine was deliberately draining the ponds at night. The Game and Fish regularly sampled the water, but had still not reported any of their findings. More than the river was starting to smell.

Finally in early September, four weeks after the Cortez meeting, the Assistant District Attorney, Wade Dillon, announced that a suit had been filed in the District Court of Dove Creek asking for a permanent injunction against Rico Argentine Mining Company to stop polluting the Dolores River. In the suit, the Game and Fish Department asked the Argentine for thirty thousand dollars in damages and costs for injury to fish life in the river. The Clean Water Coalition was elated. They'd worked within the system. They'd made the wind blow and now they were getting results.

Chapter 32

WAKE UP JOHNNY

September 11, 1964, Early Morning

Johnny had gone to bed right after dinner. After nursing Julie, Polly put her down for the night and collapsed into bed. She hadn't had a full night's sleep in two months. Something woke her in the middle of the night. Something was wrong. Her heart raced as she struggled to wake up.

"Wake up Johnny, wake up."

Julie was awakened and started to cry. Johnny had worked a double shift that day and was lost in sleep. He rubbed his eyes and tried to speak, "Polly, what's wrong."

"I don't know. Something is wrong, something smells."

He turned on the light and went to the crib to calm Julie, who had started to cough. Then he saw it. Dark fumes boiling up under the door jam and spreading across the floor, then floating up to the ceiling. The sulfur stench hit him first, then his lungs were seared by the fumes.

"Oh Polly," he grabbed Julie and pulled her against his body, gently patting her back, trying to stop the choking.

"Take her darling."

He cracked the bedroom door and then quickly shut it.

"Is it a fire?" Polly cried.

Johnny collected himself and looked around the room, "All right, stay calm, we're going outside through the window."

By now they were all choking. Johnny raised the blind, and pulled up the window, then slammed it shut.

"Damn, its coming from outside," he fought to stay calm. He saw the terror in Polly's eyes as Julie continued to choke. The plant had been down for ten days under emergency repairs and Johnny and his crew worked all day to light off the reactor. It had been a nasty mess, but they'd been lucky. A strong up-canyon wind had carried off the fumes. As night fell, he figured, the wind must have shifted. It had settled over the town.

"Put a clean diaper over her face, it's from the plant."

He pulled the mattress from the bed and set it on the floor in the far corner.

"You guys lay down over there. I've got to seal off the house."

He grabbed a blanket from the bed and quickly slipped through the door. The kitchen and front room were filled with fumes from the ceiling down to the table level, and more fumes were pouring under the front door and through a window that was cracked over the sink. His eyes and lungs were burning and he started to cough. Rolling up the blanket, he wedged it against the bottom of the door, then threw open the sink window and swung a towel in the air, fanning the fumes out the window. When he had cleared most of the room, he shut the window and raced back to the bedroom. Polly lay on the mattress, wrapped around Julie, sobbing.

"We're ok now, try to get some sleep," he whispered.

"Johnny, this is not right."

"I know, I know, I'll be back in a minute, I gotta spread the word."

He wrapped a towel around his face and was gone. He switched on the porch light and slipped outside. The light usually illuminated the houses across the street. Tonight, he couldn't see beyond his front fence. He ran down to Roy's darkened trailer and banged on the door.

"Roy, you in there? It's Johnny."

Roy came to the door in his boxers and a tee shirt. His trailer was air tight, so he hadn't noticed anything different.

"What the hell Johnny, it's the middle of the night." His head snapped back as the fumes hit him, "We got another fire?"

"No, they're still trying to get the reactor burning and the wind shifted. I gotta check on the Spitzers, why don't you call Lucy and the Fagans and get them callin' around."

And that's the way it went, a good day, the appearance of progress and then a setback, much more of an ordeal than they had figured. Johnny and Roy had passed petitions, attended meetings, and helped out in any way they could, while still going to work every day and trying to keep the plant running. But tonight was different, this was personal. Johnny got back home at 3:50 in the morning.

Polly had fallen off to sleep with Julie in her arms. They were covered in soot. Johnny watched quietly and thanked God they were ok. He knelt beside them and wiped the soot from their faces.

Polly struggled to open her eyes, "Johnny, I am really scared this time."

"I know."

He went to the window and looked out front. The wind had shifted and it had started to clear. He hadn't slept in twenty-four hours, and it took all of his effort to control his anger. It had been a horrible night.

"What can we do?"

Johnny was outraged. This was not the coy, self-aware Johnny, who chose when to scheme and when to fuss. They had crossed the line. He stared at the pool of light in front of the house and his head began to jerk, in manic fashion, from side to side. *Mess with me and I will push back, mess with my family and retribution will rain down upon you like the biblical floods.* He became a prowling, predatory carnivore, an alpha male poised for battle. Any notion of mercy, forgiveness, or fighting fare sloughed off. His eyes, the yellow hellholes of the gray wolf, cut a firestorm across the sky. Just as quickly the flame dimmed. Somehow he regained control and now he was poised, patient, and deadly. He turned back to Polly, "We'll see, but this will not happen again."

They embraced, "Johnny, we'll get through this. I love you. Please promise me...look at me Johnny... please promise me you won't do anything crazy."

"What do you mean crazy?"

"You know, you're still on parole in West Virginia."

"You're right, nothing crazy."

"I love you dearly...but I'm leaving you, if you do."

Ten minutes later, Chester phoned. Johnny and his crew needed to be down at the plant at first light. The newly installed tailings disposal line had blown and they needed to replace it. They were scheduled to work the swing shift, so they'd be pulling another double shift.

Chapter 33

THE OUTLAW GAME

September 11, 1964, Afternoon

The tailing ponds were bulldozed out of the flood plain when the plant was built. The ponds, two rows of six, were four hundred feet in circumference and held millions of gallons of red, brown, and yellow effluent. A maze of ditches and disposal lines ran down from the plant to the ponds. In theory, the solids were meant to settle to the bottom and the liquids to be diluted by the water. If everything worked properly, the water would be clean and clear at the last pond. Little effort, however, was made to regulate the output of the plant with the capacity of the ponds. From above, they looked like twelve, toilet bowls overflowing into the river.

At daybreak, Johnny and his crew arrived at the ponds. George Harrington, the shift foreman, came out of the control room and ambled across the yard. He was as big as Roy and just as mean. They were a good match. Roy was quick and agile, and George was slow and powerful. He had little tolerance for petitions, injunctions, and political action. Sure there were some risks on the job, but getting a job and keeping a job was the main thing. This had been pounded into his head by his father who had gone jobless and hungry through the Great Depression. When George got to the crew, he started barking orders. "Ok boys, here's the deal, we got an eighty foot disposal line that plugged and burst. You need to dig it up, and replace it. It runs from the valve up there below the reactor down across the yard to the first pond. That mud hole is where she broke."

A gaping hole had washed out below the line. Tailing sludge poured out of it and drained in the river.

Roy was hung-over and in a nasty mood, having spent the night passing the word about the acid fumes that had choked off the town, "So, tell me George, are you in charge of this mess?"

"I can send your ass packing, if that's what you are asking."

"Before we get into that, when do you plan on closing the line?"

George turned to Johnny, "You better muzzle your dog there."

Johnny stepped between them, "Roy, shut the line off. So, George, how deep is the rest of the line buried?"

"About three feet."

"I been told the ponds have been lined and I don't want to cut into any of that."

George smiled scornfully, "You ain't gonna run into no lining nowhere, 'cause there ain't no lining nowhere. You boys got three hours to get this fixed, so get a moving."

"Hang on," Johnny said, "Do you know why the pipe broke?"

"That ain't my department. I do what I am told and you damn well better do what you're told. Get busy."

Johnny's knuckles went white on the shovel handle, looking out on the mess, "Come on boys let get this done. We'll talk later."

"Johnny, this ain't right," Roy grumbled.

"I know, I know, and it ain't over either."

By eleven, they had replaced the disposal line and gathered for a break. Billy Gomez, a rail-thin kid from Cortez, stepped up to Johnny. He wore tight jeans with a can of Copenhagen snuff in his back pocket. He had been promoted out of the mine eight months ago and was the hardest worker on the crew, smart as a whip, but usually pretty quiet. "Johnny, can I say something?"

"Go ahead."

"That discharge line's too small, it'll plug again."

"What's the answer?"

Billy looked down to the last tailing pond, which was amber yellow, "If the system was working right, the last pond would be clear. One answer is to dig more tailing ponds, but there's no more room for that."

"Anything we can do?"

"I don't know, but they're damn sure crowding the system."

Billy stepped back, when Jim Starks joined them. He drove the ore truck from the mine, but was helping out on the day shift.

"Hey Jimmy?

"Hey Johnny, can I have a private word with you?"

"Hell Jimmy, nothing's private with this crew."

"Most of the boys on the day shift wanted me to talk to you. After what you said at the meeting the other night, most of us in the plant and up in the mine would like you to have a word with Mr. Pritchard."

"What'd ya want me to say?"

"The truth, the plant's a shit hole and it's polluting the town and the river."

Johnny looked around at his crew, "Let's have a show of hands, how many of you want me to have a word with Mr. High and Mighty?"

All hands shot up.

"Then, that's what I'll do. Thanks Jim, you boys are off 'til four, so get on home. Roy, I'll meet you at Lucy's in an hour or sooner. Bring my car and be there."

Billy came up to him as he was leaving, "A lot of us here would go with you, but we're so far in debt to the Argentine, we can't afford to lose our jobs."

"You stay put, I'll handle this."

In his gut Johnny knew it was time for a showdown, but that same gut feeling had got him in trouble before. He was pleased that for once he had gained the support of the crew before charging ahead. He didn't expect much to happen, but he was itching for a showdown.

—⁂—

Maggie knocked twice and stuck her head in the door, "George Harrington is on the other line, says it's important."

Stanley put his hand over the phone and nodded, "Look, I've got to go. If the Game and Fish are taking samples every Tuesday, then we'll drain the ponds on Tuesday evening. That will give us a week for it to flush downstream. Stay in touch."

He switched to the other line, "Go ahead, George. What? He's headed down here? No, no you stay there, I'll call the marshal."

He hung up and called Robert Tate, "Hello marshal, this is Stanley Pritchard at the Burley Building. You need to get up here as quick as you can. Well, Carnifax is on his way and he's looking for trouble."

Johnny paused in front of the Burley Building, took a deep breath, and headed up the stairs. At the landing, Maggie waved him into her office and whispered, "George just called. Stanley knows you're coming. He called the marshal. I called Myron. They'll both be here soon. Damn, ain't this a hoot? I wouldn't wait to be announced," She laughed.

Johnny glanced out Maggie's window. Myron Jones was striding up the street at full tilt. Robert "Full-Plate" Tate lurched along behind him. Johnny knocked once and walked in.

Although Stanley knew Johnny was coming, he was startled. He bolted upright and braced his arms against his desk, "You've got your nerve barging in here."

Johnny searched for the right words. He heard Myron and Robert coming up the stairs. Somehow Robert arrived first. He wheezed heavily as he struggled to tuck in his shirt.

"Marshal Tate, we've got a problem here," Stanley shouted in a cracking falsetto. His eyes widened as Myron arrived. Why is he here?

Myron sauntered over to the window and sat on the ledge, crossing his long legs.

"What are you up to Johnny?" Myron asked.

"First off, I ain't here to cause trouble."

"Marshal, this constitutes unlawful assembly, get him out of here," Stanley screeched.

Robert wiped his brow, "Now Mr. Pritchard, he's not doing, anything illegal."

Myron walked over to Stanley and winked, "Let the man say what he's got to say and then we'll get him out of here."

Stanley picked up a ruler and slapped it against his side, "All right Carnifax, make it snappy."

"This morning my crew was called in for emergency repair to the disposal line. We replaced the line, but it won't last. You're crowding the system, and nothing's getting any better." Johnny paused and looked Stanley directly in the eye, "And we're fed up with the whole mess."

Stanley slapped the ruler harder and harder against his side. He was shaking uncontrollably, "Marshal, put him in jail. He's fomenting a riot."

Marshal Tate was confused; he wasn't sure what fomenting meant.

Sensing he was trapped, Stanley walked toward the door, "Now if you will excuse me, I've got a meeting up at the plant. We are making every effort to get things right. Please close the door on your way out."

Myron blocking the door, "What Johnny's trying to say is, the boys want to work with you. Can't you give them a chance?"

Stanley squeezed past and scurried down the hall, as the marshal chased after him. Johnny and Myron shrugged in frustration. On his way out, Johnny thanked Maggie, "Good work there, sister, this is far from over."

They met Roy downstairs, stretched on the hood of the Falcon working on his ducktail.

"How'd it go?" Roy yawned.

"Not good. Let's have a drink."

Roy tossed him the car keys, "Trouble is Lucy's ain't open until three, but I got a fifth of whiskey and a quart of sweet milk."

"Myron, you want to join us?" Johnny asked.

"I believe I will."

They jumped in the car and drove way up above the plant and parked. It was about two and Johnny and Roy had a couple of hours before they had to be back at work. They leaned against the hood, sipping whiskey and chasing it down with sweet milk, a West Virginian cocktail. The afternoon breeze had shifted and the town was again wrapped in a brown cloud. From their view, they could see it all. The air was clear and the river was clean above the plant. Then came the plant, spewing fumes and acid tailings, and then the town, where the air was foul and the river was dirty.

Johnny started to bolt for home when he saw the fumes over the town, but then remembered that Polly had driven down to Dolores early this morning to stay with friends. Baby Julie had a doctor's appointment this morning.

Roy passed the whiskey and milk to Johnny, "Maybe we ought to hang ole Pritchard and his posse of lawyers from the biggest tree in town. That's how they'd handle it in a western."

Johnny nodded, "Roy's right, he's a mighty ornery fellow."

"I usually don't drink much more than a beer," Myron shrugged, "but today I think I might try a little of that sweet concoction."

Johnny handed him both bottles, "What about it Myron, you grew up around here, what do you think of Pritchard?"

"Damn that's tasty. The milk takes the edge off the whiskey, doesn't it." He slid off the hood and handed the bottles to Roy.

Roy clinked the bottles together. "It's a round trip, sin and salvation all in one go, and it beats the hell out of moonshine."

"So, what about Pritchard?"

"Well you know we've always been blessed with pretty good leaders, people like Swickheimer, the Pellets, C.T. Van Winkle, and Ed Baer. Now Ed, he was a good one."

"Was he the one with the hook like a pirate?" Johnny asked.

"That's him. His shirt sleeve got caught in the rolls at the Pro Patria Mill, damn near swallowed him up. The miners loved ole Ed. They used to say, 'If it's all right with Ed Baer, it's all right with the rest of us.' He ran the mines and was Country Commissioner for years."

"So, what about Pritchard?" Roy persisted.

"None of these guys were angels. They were strong-willed and had a lot of power. But for the most part, they were good-hearted. They cared about the people and the community."

"I guess you don't want to talk about Pritchard," Roy snapped.

"Let's set Pritchard aside for a moment. Yeah, he is part of the problem, but it's bigger than him. If we ran him out of town today, another one just like him would pop up overnight like a mushroom."

Myron wiped the milk from his mustache, got down on his haunches and started sharpening the end of a stick.

Roy still wanted a public hanging, and was unhappy with what he was hearing, "So Myron, you're the mayor. Do ya think this injunction business will work?"

"I think Pritchard and his law dogs will find a way around it. Ya see when you are enjoined it means you are legally bound to stop doing what you're doing, once and for all. So if he was serious about this, he would have to shut down the plant, fix things properly, and then have the folks up from the Game and Fish to verify it was fixed."

Johnny rolled his eyes, "Let me get this straight, polluting the river is against the law. So they broke the law there. And if an injunction is the law, then they're breaking the law again."

"Come on, Johnny," Roy slammed his hand on the hood. "We've been to this movie before. Lawyers are educated outlaws, and as my daddy always said, 'when the educated outlaws have the upper hand, you need to get into the outlawing business.'"

"Yeah, but we're the ones who usually go to jail," Johnny grinned and then felt guilty. He promised Polly he wouldn't do anything crazy.

They sat there for a while, sipping quietly trying to work out a plan. Johnny was haunted by what he had learned from the Navajos. Norman's words kept playing through his head, 'When the land is sick, the people are sick.' He looked down at the plant and what he could see of the town. Norman damn well got it right. And he thought, you could turn that adage on its end as well, when the people are sick, like a lot of assholes around here, the land is sick. At that, Johnny's head snapped up, "You boys believe in dreams?"

"What ya got in mind?" Myron stood up and stretched his legs.

"When Norman was dying he told me about the fire in the meadow and the fire on the mountain, an Indian thing. Said it came to him in a vision. You ever hear that one?"

"Come on Johnny, get to it," Roy fussed, reaching for the bottles.

"Norman said the fire in the valley feeds your body and the fire on the mountain feeds your soul, said you had to keep both fires burning. I've been having some mighty bad dreams about all of this."

Roy scratched his head, "Sound like you'd be doing a lot of chopping and a lot of running."

"You got to be careful with them Indian visions," Myron warned. "They are tricky, like riddles. Maybe it's about making the right choice when you can't do both. Hard to say. Makes you think, don't it?"

"Anyway, where do we go from here, Mister Mayor?" Johnny asked.

Myron got down on his haunches and cleared a small patch of ground, "Since we talked on the Fourth of July, I've given this a lot of thought. I am still wrestling with how things went bad here in Rico and down on the Colorado Plateau. I figure there are several moving parts in this equation. I mentioned two of them back on the Fourth, the opportunity for obscene profit and the faceless power of the big mining corporations like the Argentine."

"Other than your mines, the Argentine owns near every mine around Rico, don't they?" Johnny asked.

"Yeah, that's right."

"And they are making a fortune selling acid to the uranium plants," Roy added.

"So boys, I take it we agree on these two factors, corporate power and obscene profit." Myron took a stake he been sharpening and drew two lines in the dirt.

"Why are you drawing lines in the dirt?" Johnny asked. "Last time you staked them. I liked that part."

"It's a small thing. You see a stake is kind of permanent. It stands alone. You draw two lines side by side and they interact, you know, one influences the other. Anyway, I need your help. What else is involved?"

Roy was on his haunches now running his index finger down the two lines, asking himself if there were other factors. Then it came to him, "You got all these agencies like the Public Health Service, the Game and Fish Department, and District Attorney; they don't seem to be doing much of a job for us."

Myron smiled broadly, "You're right on that, I don't know if they've been bought off or if they just ain't worth a hoot. Let's call it poor regulatory oversight. No, let's just call it inept, they really are inept, let's call it inept oversight."

Johnny grabbed Myron's stick and drew a third line in the dirt.

At this point Roy's legs started to shake, "I'd like to get back to Pritchard, poor leadership has got to be one of the factors."

"You're right about that Roy, but let me tell you what I've been thinking. At first, I figured it was all about poor leadership, what I call unprincipled leadership. But I have had a change of heart. Here's the way I am thinking now. It's the first three factors: corporate power, inept oversight, and the opportunity for obscene profit that corrupts the leader. So, I figure old Stanley ain't the villain we'd like to make him, he's just a pawn in the game."

"That is some mighty fancy thinking Myron," Johnny said.

Roy rocked back and forth on his heels. He was tired of all the talk, "Okay, fine, but I still don't like the man." He grabbed the stick from Johnny and made a fourth line in the dirt.

"I don't like him much either." Myron agreed. "So boys, I think we're all coming out about the same on this."

They passed the whiskey and milk back around, and stared down at the four lines in the dirt.

"I call this, The Four Pillars of Greed. You get these four factors working together and you end up with greed. I mean like every single time. It's like a mathematical formula. So, that's what we're up against boys."

"The Four Pillars of Greed, that's mighty slick. Where'd you get the name?" Roy asked.

"That part just came to me now. I do some of my best thinking when I am drinking."

Roy jumped up and started dancing around the car, "Hot damn, Johnny, Myron's an outlaw too."

Johnny frowned, "Myron, if you're right about this greed business, it's downright depressing, that's a whole lot to deal with."

They leaned back against the hood and drank silently. By three-thirty they were ready for anything.

Then Roy got a twinkle in his eye, "I think I got it." He jumped off the hood, picked up Myron's stick, and drew an arched roof over the four pillars, "What da ya think?"

"It's a bank or a courthouse?" Johnny asked.

Roy slashed a line across the four pillars, "Na, it's the Temple of Greed, it's what Norman called the bad wind. Now help me here. Norman's last act was to take his records to the PHS, am I right?"

"That's right, we reckon he died coming over Lizard Head Pass," Johnny replied.

Roy danced around the Temple of Greed, flapping his arms up and down like he wanted to fly, "I think I figured out the riddle. Let me ask you this, which fire was Norman building when he took the records to the PHS? Was he doing things right, or was he doing the right thing? Was he feeding his body or feeding his soul?"

Myron pulled on his mustache in amazement, "Roy, that's apocalyptic, an apocalyptic revelation."

Johnny looked at Roy suspiciously, "How'd you figure that out?"

"Trish and I have been writing them songs, you know, and we've been doing a lot of thinking, thinking about doing what's right, and it all just came together for me. Do you want to work for the man down in the muck or do you want to fly with the eagles? I can see it all now, we've got to feed the fire on the mountain, and we've got to do what's right."

Johnny looked down at the Temple of Greed in amazement, "You'd better lay off the whiskey, you're starting to sound like a evangelist. So Roy, since you are seeing all this so clearly, what's the next step?"

"We gotta to bring down the Temple of Greed."

"Hum? How you gonna do that?" Myron asked.

"Here's the deal, we drive down to the plant, and we tell 'um, fix it or we'll shut her down, and by golly if they don't fix it, we'll shut her down."

Myron rubbed his temples, "Why would they want to fix it?"

Johnny starred up at the summit of Blackhawk Mountain for the longest time. Snow chutes ran recklessly down its steep face. The spruce and the cottonwoods seemed to be marching toward the acid plant. Maybe it was the whiskey, but he felt the mountain was on his side, urging him on. He jumped off the hood and slid behind the wheel, "Roy, that's pure genius. Come on boys we got work to do."

They piled in the car and roared off down the hill.

Roy was about to bust, "What are we up to?"

"First, we need our equalizers."

"Our what?"

"First we get our guns, and then we go to the plant."

Myron's head popped up, "Boys, you better take me home."

"That's fine, you've done enough there Mister Mayor." Johnny chuckled.

Chapter 34

TAKING ON THE MAN

September 11, 1964, 4 PM

Johnny dropped Roy off at his trailer and headed home. Sulfur fumes had crept back into the kitchen. He pulled his Stevens, twelve gauge shotgun from under the bed and stuffed a handful of shells in his pocket. He did his best to seal off the house and then picked up Roy. They drove down the hill to the Argentine property line and parked inside the entrance gate. It was windy and dark clouds rumbled on the peaks. Johnny pulled a padlock from his glove box, opened it and tossed the key in the backseat. Grinning broadly he swung the heavy metal gate closed, stretched the chain around it, and snapped his padlock shut ahead of the others. No one was getting in or out without a bolt cutter.

Roy was goofy with excitement. He rubbed an oil rag up and down the barrel of his British Enfield 303. This bolt-action, repeating rifle was the main firearm used by the military forces of the British Empire. It was a great gun for hunting deer or starting a war.

"Johnny, I ain't ever held up no acid plant."

"Me neither, let's call it on-the-job training."

They parked outside the control room and loaded their guns. With his 303 in hand, Roy popped out of the car.

"You let me do the talking and let's leave the equalizers here for now."

The traffic had already started to back up at the gate. Huge ore trucks coming into the plant and empty acid tankers returning for another load were lined up outside the gate.

A loaded tanker heading down to the Uravan Mill was locked inside. Most of the drivers figured out what was going on and made little effort to open the gate. Even Okie Tom Barlow was a bit subdued.

George Harrington, the day foreman, frowned at the boys as they walked into the control room. His crew stood behind masking their smiles.

"What the hell are you up to?"

"Well George, I'll give it to you straight," Johnny spoke in a calm measured tone, "we're fed up with the working conditions in the plant, and we're fed up with the mess it's making around town."

"Look, why don't you hillbillies hit the road."

"You being the foreman and all, we figured we'd put you in the driver's seat. You've got a decision to make, either fix this place or we're shutting her down."

The crew gave way as Roy moved to the control panel and found the in-feed valve.

"We've done all the fixing we're gonna do around here."

"Well George, if that's your final word, we're shutting her down." He nodded to Roy, who closed the in-feed to the reactor.

George lunged for the control panel, but stopped in his tracks, as Roy swung a two-foot spanner wrench in the air, "Like he said big shot, you made the call."

The in-feed valve controlled the flow of pyrite from the crusher to the reactor. If it was cut off for more than twenty minutes, the reactor would freeze up causing massive damage to the whole system. Johnny was running a bit of a bluff. If they knew what was at stake, he figured they'd be more willing to fix things.

"The rest of you boys go on home," Johnny said. "Roy, walk them down to the road."

From the road below the plant, Roy could see the north end of town. By now the word had spread and a crowd had gathered at the gate. The town marshal, Robert Tate, was having fainting spells, but had called the Dolores County Sheriff, Bad-Bob Johnson, from Dove Creek and the Assistant District Attorney, Wade Dillon, from Cortez. A State Highway Patrolman had just arrived, siren screaming and red lights flashing.

Roy was as giddy as a teenager after his first six pack of beer. He ran back up to the control room, "Hey Johnny, we're going to war! The Highway Patrol just arrived and I am sure they've called in every cop in the county."

"Settle down Roy, we ain't going to war. We just want to talk some sense into these folks."

Several minutes later, Chester barged in the control room. He'd been up at the Saint Louis Mine. "What in hell is going on?

Johnny stepped up, "Like I told George, we're fed up with the working conditions at the plant, and we are fed up with the mess it's making around town. You boys need to fix things or we are going to shut her down. George there, refused to fix things, so we're shutting her down."

"We're not putting another penny into this operation. You boys better get out of here, now."

"Chester, you know things are getting worse, and it won't cost you much to get it right."

"The hell you say," he walked over to the phone, "I am calling Pritchard."

Johnny took Roy's spot in front of the in-feed valve and whispered to him, "Go get the equalizers."

Alarm bells were wailing throughout the plant, and the control panel was a blaze of flashing red lights. Moments later Roy ducked back in the room and handed Johnny his shotgun. He cradled his 303 in his arm.

Down at the Town Hall, Paul Spitzer, the Justice of the Peace and Johnny's brother in law, made a call to Bad-Bob Johnson, "Hello, is Bob there? This is Paul Spizer in Rico."

The dispatcher told Paul that Bob was on his way.

"Ok fine, well can you get him on the radio? Good, tell him I know the boys that have taken over the plant. They're good boys just had a little too much to drink. Tell Bob if he's got to shoot them, to go ahead, but go for a leg. Good...thanks...you bet...thanks."

Back in the control room, Chester had just hung up, "Well first off Pritchard said you're both fired. The plant is surrounded. They know you have guns and they're coming in shooting."

Johnny sat on the desk with the shotgun next to him, "So I take it, that you mighty captains of industry are going to let the reactor seize up on you."

"No, I am saying that within minutes you'll be in jail," Chester barked.

Johnny looked at the brass plate on the top of the control panel. It read Leonard-Monsanto. They had built the plant. They couldn't be happy with the way it was being run. Maybe they'd shut her down as well. He struggled to remain calm, "Look, all we're asking is that you clean up this mess. You're the ones that are breaking the law. Just say the word, and we'll cut the in-feed back on, and everyone comes out a winner."

And so it went, back and forth, more calls to Pritchard, more arguing, more ultimatums on both sides, a standoff. By now the State Troopers, the County Sheriff, the Assistant District Attorney and the Town Marshall had assembled at the gate and were developing a plan to retake the plant. Chester was back on the phone. Johnny left Roy in charge and stepped out to check the reactor. Sure enough the plant was going cold. He sensed that Chester wanted to negotiate, but Pritchard wouldn't budge. Afterwards, he heard that Pritchard had locked himself in his office and called Governor Love, trying to get him to call in the National Guard. Johnny taped the temperature gauge on the reactor. It was dropping quickly; a few more minutes and the molten pyrite inside would set up like concrete. He wondered how Polly was handling all this. He knew she would be scared, but probably no more than when they were suffocating in sulfur fumes.

It looked like he'd have to take it right to the edge, to have any chance of changing things. But they'd taken it to the edge, and the Argentine wasn't flinching. He looked back at Blackhawk Mountain, a jagged silhouette of snowcapped peaks in the gathering darkness. Ribbons of spindrift swirled off a high cornice like an infantry banner leading an army of the night. It gave him a comforting sense of permanence as did Shiprock and Lizard Head Peak. Slowly wearing away, but immutable by human standards. He felt part of it, maybe that was *hozho*. The wind howled down off the mountain and banged the tin siding on the ore house like a drum roll, a long attack drum roll. That must be a sign, Johnny concluded, give it one more try.

When he got back to the control room, Chester was still on the phone. Johnny grabbed the receiver and hung it up, then sat back on the desk, "Chester, let's talk. I am only going to say this one more time."

"I am through talking." Chester said shaking with rage.

"Then try and listen. You've got about five minutes to save that reactor out there. All we're asking is that you stop breaking the law."

Chester bolted suddenly for the door. Johnny grabbed the shotgun in one hand and pulled the trigger. The handle of the door blew off. Buckshot rattled around the control room like rocks in a tin can. Chester and George hit the floor. Roy cackled like a crow in a cornfield. No one was hurt, but the door now swung freely.

The negotiations were over, and a cold wind raced through Johnny veins. He was now the iceman, "Sit down, both of you, and give me straight answers."

—∞—

The shotgun blast echoed down into town. Everyone heard it and everyone froze in their tracks. One shot. They waited. None followed. The stakes had just gotten deadly serious. By now Bad-Bob had taken charge and the plant was surrounded. He'd cut the chain off the gate, drove up near the tailing ponds in his pickup, and blocked the road. His 30 odd 06 Springfield rifle laid across the hood.

Polly had returned from her doctor's appointment and joined Maggie at Lucy's. They huddled at the nook across from the bar, nervously sipping tea. As she heard the gunshot Polly bit her bottom lip, tears rolled down her cheeks, "I know he's crazy, but he's a good man."

Maggie took Julie in her arms, "Don't you worry. Johnny is at his best when things get crazy."

Lucy refilled her cup, "Maggie's right, we've got to stay strong."

Polly's eyes narrowed as she wiped the tears from her cheek. "Yeah, I'll stay strong, but he knows how I feel. If he goes to jail, I'm gone."

—⋙—

The arguments continued up at the control room. Roy had got on the phone and called his old gang, the Ridge Runners, in Chicago and a carload of them were headed for Colorado. They too had been drinking and figured if they drove all night they could make it to Rico in the morning.

Johnny had been tempted to call the whole thing off, but he was reminded of what Chee had once told him. According to Indian lore, the Anasazis of Chaco Canyon had broken their pottery when under siege to keep it from invaders and as a sacrifice to the gods. Maybe they needed to break a little pottery. He missed Chee, but he was off fighting the bad winds on the reservation.

Minutes later Johnny turned to Chester and George, "All right, it's over. You made the decision to shut her down, and she's down for good. Go ahead, get out of here."

The control panel was dark by now and the plant had gone still. Both the pressure and the temperature gauges were pegged at zero.

No fumes came out of the stacks and the sky was clear for the first time in ages.

Roy walked outside and scratched his head, "What do we do now?"

Johnny laughed, "I don't know, we damn sure shut her down though, didn't we?"

"We damn sure did. But right now, I got me one hell of a headache, and I am starving."

"That's called a hangover. You stay here and let me scout things out."

Johnny's mouth was dry and his head was pounding. He wasn't doing his best thinking anymore. He jumped in his car and headed for town. As he came round the curve below the tailing ponds, a red pickup truck was parked across the road. Someone yelled stop. By the time he heard the shot, the front seat cushion had blown apart. He skidded around and headed back up the road. He wasn't hit, but he was badly shaken. The bullet had somehow gone through the left fender and on through the firewall into the seat cushion under him. An inch higher and he would have lost his manhood.

Roy ran out of the control room as he drove up, "What in hell was that?"

"The bastards tried to kill me. They're shooting to kill."

Just then the tailing ponds were splattered by a burst of rifle fire. A second burst bounced off the heat exchanger and the absorbing tower.

"Whoa! They're a coming. It's okay, we've got enough ammo to hold 'um off 'til dark. The Ridge Runners will be here in the morning."

"No sir, we did what we came to do," Johnny declared. "No more guns, no more shooting. It they want us, they can come and get us."

Another burst riddled the galley and feeder tower just above them.

"Damn, what's the plan!"

"Come on let's get a cup of coffee and wait."

They sat and sipped their coffee and waited for over an hour.

"Johnny, I been thinking. In West Virginia, we did pretty well. We got them to do the roof bracing after all them cave-ins and we got them to turn on the ventilation fans after the methane gas killed a couple of our boys."

"That's true, what's your point?"

"We also went to jail for those things."

"Yes, we did." A pang of guilt ran through Johnny. He'd promised Polly he would do nothing crazy today and damned if he hadn't broken his promise.

"My point is, today we stopped the acid rain and the river pollution here in Rico, but we'll probably go to jail again."

"Can you shut up about going to jail?"

"We came down here to topple the Temple of Greed and I think we failed. Sure we shut her down, but I don't see how we have changed the Four Pillars. How is what we did here today going to change corporate power, inept oversight, obscene profit or unprincipled leadership. I guess I am asking you, how do you prevent this from happening again?"

"First off, we did shut her down, and she may be down for good. Hell Roy, no one said we were going to save the world today. But we damn sure didn't sit on our butts and let them get away with it. I think you get moody when you haven't eaten. As for the rest of it, you don't have to win every round to win a fight. Be happy, we won this round and we kept our promise to Norman. It has been a good day. No use gettin' killed."

"Maybe you're right. I am walking to town for a sandwich. Do you think Polly will leave you?"

"I hope not, but that is none of your business, pecker head. Walk slow and stay out in the open going to town."

"I'll follow in five minutes. If…"

"If what?"

Johnny grinned, "If they don't shoot your bony ass."

"That ain't funny. Shit, I 'm puttun' my handkerchief on a stick, just like they do in the movies."

By six o'clock Johnny and Roy were standing before Paul Spitzer, the Justice of the Peace (JP), in the packed Town Hall courtroom. This was high drama for a little town. Some came for a public hanging and others for a celebration. After they were taken into custody, the sheriff and District Attorney questioned the boys. Bad-Bob smiled broadly and apologized for the near miss, said he was aiming for the left front tire, but shot a little high. Chester signed a complaint charging them with assault with a deadly weapon. The criminal complaint said: *willful and unlawful assault with a present ability to commit a violent injury on the person of him, Chester Ratliff.*

Johnny and Roy stood before the bench in handcuffs, as the JP read the counts. Most people in the room were surprised by the charges. Johnny looked around the room. He caught Chester's eye for an instant and he could swear Chester winked at him. Everyone in town was there, but Polly. His heart was pounding like a jackhammer. *She finally left me.*

"All right boys, you are hereby ordered to appear at a preliminary hearing at the Dove Creek County Courthouse on September 21st. I am setting an appearance bond of one thousand dollars on each of you. And, unless you can come up with the bond, the sheriff will take you down to jail in Dove Creek tonight."

A buzz went through the crowd. No one had that kind of money in Rico, well very few. As Paul was about to pound his gavel, Myron Jones, who was leaning against the wall at the back of the court room raised his hand.

"Yes, Myron."

"I got 'um covered."

Paul smiled, "Ok boys, you can go home to your families for supper. Make sure you're in Dove Creek on the 21st."

Johnny smiled lamely to Paul, "You got room in that jail?"

Johnny tiptoed onto the porch. His head was pounding and his mouth was sour. Waves of emotion washed through him—guilt, remorse, and a muted sense of triumph. Polly's suitcase and a smaller one for Julie were on the porch. So they were leaving. He peaked inside. Polly sat stoically on the couch rocking Julie. Storm lay at her feet. Johnny lost his balance as he bent over to pull off his boots. He caught himself against the wall.

She looked up as he stumbled through the door. "Look Julie, it's daddy. He's had a busssssy day."

Johnny winced as he crumbled into the armchair.

"Yes, daddy's been a busy boy. He got to drinking and then he got into a gunfight, and now he's going to jail."

"Ah, come on Polly, there's more to it than that."

Polly's eyes were red and her cheeks were swollen. Julie squirmed in her arms, starting to fuss. She walked to the bedroom door and stopped, "Johnny, I have had all of this I am gonna take."

Johnny gripped the arms of the chair until his hands cramped. Storm sat near the couch but wouldn't come to him. Polly disappeared into the bedroom. Her hat and coat and Julie's hung over a chair at the kitchen table. They were leaving him? *This can't be happening.* "At least, hear me out before you go."

She reappeared and stood rigidly at the doorway with her arms locked across her body. She was laughing and crying at the same time. "Look at me. I'm a mess. I am happy you are alive and disappointed they didn't shoot you. You got no right to mess with my feelings like that."

"Polly let me explain."

"Johnny, shut up." She caught him looking at the hats and coats on the kitchen table and started laughing again.

"I never met a man that did more things wrong for...for the right reasons."

"Just say you'll stay, Polly."

"I'm not leaving you, but I damn sure ought to."

"What about the suitcases on the porch?"

"I just haven't unpacked them."

"Thank God, come on Polly, let me explain."

"I will never accept the guns and the shooting. But, come here ya big lug."

They embraced. Polly's forgiveness washed him clean. He was renewed, redeemed, and ready to start over.

"Did you figure out the business of the two fires?"

"We did. Myron and Roy helped me with it. Myron said it was a like an Indian riddle. Us white guys are not good at that. He said it was about making the right choice, when you can't do both. Roy said we have to do the right thing, not just do things right. I think he got that from a billboard. But it sounds true. So we followed Norman's example and fed our soul."

Polly stood up and looked him in the eye, "You'll never be a saint, but I think most of what you did was right. Good luck to you in explaining all that to a judge. So, let me ask you one more question before you go to jail."

"That ain't funny."

"What was it like shutting down an acid plant?"

Johnny's eyes lit up and he grabbed her hand, "Well darlin', nothing beats being your husband, but I would put it way above fighting and football."

"It's a hell of a country where you have to break the law to do the right thing. We damn sure did everything we could within the law, to make things right," Polly fussed. "But what the hell, as outlaws go, you ain't bad."

They embraced as Storm bumped up against their side.

"You still love me then?"

"You're damn' tooting."

"Then help me pack the car. We gotta be moving along."

THE END

Epilogue

"Public sentiment seemed to be with the boys."
—Hart's Stuff, Dolores Star, Sept. 1964

When trust is high and people share common values on protecting public land, simple communication can solve many problems. The more you network and make your case, the better you do. It is also a naïve strategy when dealing with outlaws. But let us take a moment to celebrate the Coalition for Clean Water (CCW) in Rico. They did more than most colonial outposts in the high country in using the existing legal system of due process—petitions, injunctions, and appeals. Trouble was, they were dealing with educated outlaws.

That's where Johnny and Roy were helpful. They'd learned the outlaw game and demonstrated a fascinating anomaly of Frontier Justice:

Educated Outlaws on occasion beget Benevolent Outlaws.

So, for those that want to make the world a better place to live, take heed. It seems as though you need both—benevolent coalitions and benevolent outlaws.

EPILOGUE

The acid plant went cold on 9/11/64. Now, that's a sign. Jobs were lost, but most people knew it was time to move on, or find another way to stay. Johnny and Roy were sentenced to six months in jail and fined two hundred dollars each. The sentence was suspended and for the next week or so, folks from up and down the Dolores River bought the boys drinks and gave them traveling money. After the shoot-out, the plant never reopened. Within weeks the river was running clean and clear, though they are still cleaning up the mess forty-six years later. Through some legal sleight of hand, you just can't trust an educated outlaw, the injunction against the Rico Argentine was dropped the following May.

Johnny with his Stevens 12 gauge
at a coal tipple near his home, 2009.

Johnny and his family moved on, reluctant nomads. They ended up in Hopewell, Virginia where Johnny worked for Firestone Tires until he retired. They raised three wonderful children, John Jr., Julie and Lisa, all of whom graduated from college and have done very well. Polly and Johnny drifted apart after the children left home. Today, Johnny lives alone deep in a hollow near Rainelle, West Virginia on a patch of mined-out land.

I have joined Johnny on several morning hikes to the top of the mountains near his home. The skyline captures the twenty-first century struggle for energy independence. To the north, coal tipples scar the hillsides—dirty, non-renewable energy. To the south, hundreds of wind generators stand tall on the ridgeline. Enormous vanes turn slowly on three hundred foot towers—clean, renewable energy. Kayford, WV, the largest mountaintop removal operation in Appalachia, lies seventy miles to the west. And a few hollows to the south, is Montcoal, WV.

And so it goes. The dark side of unbridled capitalism that Rico experienced during the Cold War is still a part of the Montcoal mine disaster, and the BP Gulf oil spill, and the Wall Street meltdown, and so on.

In the late sixties, the demand for uranium peaked on the Western Slope of Colorado, as did the demand for sulfuric acid, but not without taking its share of casualties. On January 1, 1969, Willard Wirtz, the Secretary of Labor, enacted regulations that prohibited uranium mines or mills from operating if the radon gas exceeded 0.3 working levels. Of course, this was too late for many of the mine workers. Based on a study concluded in the year 2000, one thousand five hundred and ninety-five (1,595) white and Navajo uranium miners died from various lung diseases on the Colorado Plateau (NIOSH Uranium Miner's Health Study).

Not until October 15, 1990 was the Radiation Exposure Compensation Act (RECA) signed into law, based on the efforts of Harry Tome, Timothy Benally and the Widows of Cove. Mine workers who had been exposed to two hundred or more working level/months of radiation from January 1, 1947 and ending in December 31, 1971, and who had contracted lung cancer or other serious respiratory disease would be awarded $1,000. If the miner died, the award went to their widow or children. The struggle had just begun for many of the Navajo widows, few of whom had marriage licenses.

The tragedy, of course, is that all of this could have been prevented with proper ventilation and a few inexpensive safety precautions. The Atomic Energy Commission and the U.S Public Health Service were aware of the deadly connection between radioactivity and lung cancer since 1944. That year, a report by the National Cancer Institute based on numerous studies of pitchblende miners in Europe was made public.(Radioactivity and Lung Cancer: A Critical Review of Lung Cancer in the Mines of Schneeberg and Joachimsthal). Pitchblende is the chief ore-mineral source of uranium. The European studies found that seventy-five percent of the deaths among miners were from lung cancer, what the Germans called *Bergkrankheit*, or mountain disease. The report was published by the U.S Public Health Service and then stuck in a drawer.

Acid Plant, 1978, fourteen years after the plant closed.
The trees in the foreground where killed by acid rain.

In regard to clean air and water, there were a variety of federal laws passed to protect the environment in the 1960s. The Clean Air Act of 1963 attempted to clear the skies of harmful pollutants like acid rain and radioactive contamination, and in 1966 the Water Control Act aimed to improve water quality. And then in 1969, the National Environmental Protection Policy Act was signed into law, which paved the way for the Environmental Protection Agency (EPA) which was initially limited to providing technical assistance to state and local governments.

Finally in 1972, the Federal Water Pollution Control Act (FWPCA) was signed into law which enabled the EPA to file civil actions and set daily penalties for polluters in violation of state standards. The laws were now on the books and the EPA had the power to enforce them.

Typical of many areas where intensive hard rock mining had occurred, Rico was an environmental casualty.

Acid mine drainage, acidic water, occurs when water (rainfall, streams or snow melt) drains through mines with exposed metal sulfides, often pyrite. Rico has plenty of water runoff and plenty of mines with metal sulfides. The acid drainage from Blaine Tunnel and various other mine adits on Silver Creek contribute to cadmium and zinc loads that exceed the standard load to protect aquatic life. Acid drainage from the Saint Louis Tunnel (directly above the acid plant) and breaching tailing ponds (below the acid plant) have also exceeded standard, and have added to the water contamination. Over the years, efforts have been made to improve the water quality with some success. Finally, last year (2010), the EPA developed a plan and directed the Atlantic Richfield Company (ARCO), the primary responsible party, to begin the clean-up of the Saint Louis Tunnel. This will involve building a water treatment plant at the Saint Louis adit to treat acid drainage, dredging the first of eighteen tailing ponds, and building a repository to place the contaminated sludge. This is good news, but a longtime coming.

ACKNOWLEDGEMENTS

The first time I heard an account of the shoot-out at the acid plant was on my mother's deck during the summer of 2005 in Rico. My mom, Rhea Curran, was ninety years old. We were enjoying a bottle of wine with Carole Rychtarik, our neighbor, and watching the sunset over Sandstone Mountain. Carol shared bits and pieces of the story that she had experienced or heard about while living in Rico. We enjoyed the tale, thinking it a bit tall, but up to Wild West standards. Later, at the Enterprise Bar, we heard another version of the tale. As we headed home, mom asked me if I would research the acid plant incident and write it up so it wouldn't get lost. And that is how this book got started. Mom died in 2006 but I still hear her asking, "How are you doing on that book, son?"

"Soon, Mom, soon."

So, I went about phoning and interviewing old timers around Rico. For several months, I found no one who could remember when the incident occurred or the names of the two West Virginia roughnecks that shot-up and shut-down the plant. Then, we formed a research posse: Sue Eleison, Carol Rychtarik, Glenn Baer, and Jim Starks all long time residents of Rico were charter members. Marilyn and Jim Skelley, two friends from Telluride, completed the posse. We all went to work pouring over old newspapers and asking around, but no one could remember the day or the year it happened.

Finally, Marilyn Skelley, bless her, found an article about the incident in the Telluride Times. It happened on September 11, 1964 and the names of the two boys who shut down the plant were: John Carnifax and Roy Loudermilk. The posse has continued to provide excellent support over the years.

Glenn Baer should be dubbed the Dean of Rico History. We are constantly on the phone. He was enormously helpful in recalling the Acid Plant era in Rico as was Jim Starks who had great recall on both frightening and funny stories. Marlene and Ken Hazen joined the posse later and have done a great job tracking down photographs, addresses of old timers, and historical facts. My brother, Mike Curran, was also drafted into the posse. Thank you all for your continuous help and support.

The more I researched the health and safety problems caused by the acid plant, the more I realized how closely linked they were to the health and safety problems caused by uranium mining, on the Colorado Plateau and the Navajo Reservation. For their counsel in helping me understanding the Navajo culture and the mining experience, I first want to thank Jim and Grace McNeley, both are professors at the Dine College on the reservation. Grace spent several years in Rico as a school girl. We have met several times and exchanged various versions of the manuscript. Next, I want to thank Timothy Benally who showed me around Cove, Arizona and helped me interview, David John and Paul John Nez, uranium miners from Cove. Timothy was the Director of the Navajo Uranium Workers Program and played a major role in achieving federal compensation for the families of Navajo miners who died of lung cancer from radioactive exposure. He also worked in the mines of Rico and in Cove. Harry Walters, an elder from the Cove Chapter, also helped me understand the Navajo Way and the challenges of the Navajo assimilating into the white man's world. In Rico, I want to thank Linda and Norman Yellowman for sharing their experiences of living in Rico. Linda is the Town Clerk and has been very helpful in providing access to the minutes of the Town Board meetings during the acid plant era.

After almost a year of researching the book, I finally drove up to Rainelle, WV and visited John Carnifax for several days. We spent hours talking, hiking, and visiting old coal camps. Like the Navajo experience, understanding the history of coal mining in WV and how John and Roy grew up was invaluable. Jeanette Thomas, a neighbor of John's, has been a critical link to John. I sincerely thank you both.

Dr. Jerry Griebel of Rico and Dr. Mike Maffett of Atlanta have patiently responded to my many questions on the relationship between radioactivity and lung cancer. Thank you Doctors.

ACKNOWLEDGEMENTS

Jenifer Stark and Becky Levy were involved in developing the Rico Master Plan and have been heavily involved in Voluntary Clean Up projects with the EPA, the Town of Rico and Atlantic Richfield. Thank you for helping me understand the issues. Phil Egidi and Mark Rudolph, from the Colorado Department of Public Health and Environment provided information on past and current environmental issues. Ron Evers who was involved in tearing down the acid plant provided valuable photographs.

Maggie Matzick, who recently passed away, was as enthusiastic as a teenager when recounting her fond memories of Rico. Both her parents, Helen Hicks and George E. Hicks, were Superior Court Judges for Dolores County. Both were pillars of the community.

I have conducted detailed interviews with the following people: Bob Johnson, Dorothy Spitzer, Dennis and Verbil Swank, Chet Towns, Audrey Goucher, Jim Fahrion (Lucy's son), Jay & Mary Lou Milstead, Deanna and Val Truelson, Chet Townes, Max Baer, Erin Johnson, Felix Snow, Jim Barren, Todd Jones (Myron's stepson), Kay Crane, Steve and Mary Lou Hudson, Erin Johnson, Barbara Betts, Mike England, and Larry Fitzwater. Thank all of you for your time and your insights.

Special thanks also to Lauren Bloemsma, Director of the Telluride Historical Museum, to Peter Kenworthy, Director of Mountainfilm in Telluride, Sarah Landeryou from the Telluride Library, TJ Holmes the Editor of the *Dolores Star,* and Eunice Kahn, Archivist at the Navajo Nation Museum in Window Rock, Arizona.

For their patients and hard work in proofing the manuscript I want to thank: Patrick Scott, Dr. Vernon Sanders, Judy and Jeff Godard. Thanks also to Dell McCoy who wrote the wonderful Rio Grande and Southern Series for his encouragement and willingness to permit the use of photographs from his books. And a very special thanks to Jill Murphy Long and Caroline Metzler, for their excellent editing and graphic design, and to Thomas Clyde for his advice on historical fiction.

Two books that were invaluable in understanding uranium mining on the Colorado Plateau were: *If You Poison Us,* by Peter H. Eichstaedt and T*he Navajo People and Uranium Mining,* edited by Doug Brugge, Timothy Benally, and Esther Yazzie-Lewis.

And, of course, a heartfelt thanks to my mother who got me interested in the book, and to my lovely wife, Joan, for all the chores I didn't do while writing and for all the moments when I was there, but really not there, reworking the book in my head.

Photo Credits

Page	Source	Description
cover	Caroline Metzler of Coyote Creative	Montage: acid plant, Johnny, and Navajo Women *Courtesy of Navajo Nation Museum*
(i)	Telluride Historical Museum	Ore specimen, SKU 1995-119
10	2005 Sandia Software	Lizard Head Peak
14	Courtesy, History Colorado Collection	Rotary snow plow, CHS.X9550
17	www.wvgenweb.org/wvcoal/p24.html	MacAlpin Coal Camp
29	Rachel Hardwick, Charlie Engle Collection	Rico Mercantile, 1890s
34	Lucy Fahrion Family, Dell A. McCoy Collection	Enterprise Bar & Lucy superimposed
42	Todd Jones Collection	Myron Jones, 1980
48	Rachel Hardwick, Charlie Engle Collection	Prospector heading for his claim
58	Glenn Baer Collection	Glenn Bear, operating the three drum slushier.
70	Monte Ballough Photo—Larry Pleasant Collection	Women's Club out for a ride after church
79	Glenn Baer Collection	Rico Acid Plant 1960s
83	Glenn Baer Collection	Control Room, Edgar Branson left, Jack Darnell right
96	Telluride Historical Museum, Pro-Patria Aerial Tram	Aerial Tram from Pro-Patria Mine to Mill, 1912
114	Glenn Baer Collection	Ore House the morning after fire
130	Fort Lewis College Center for Southwest Studies, P001, V-01-08	Roof bolting
136	Courtesy, History Colorado Collection, Bob Zellers	Rico Argentine mine: dumping mine tailings into Silver Creek

PHOTO CREDITS

Page	Source	Description
140	Dell A. McCoy Collection	Rico Town Hall
171	Courtesy, History Colorado Collection Bob Zellers CHS.X5525	Shamrock Mine above U.S. Vanadium Mill in Uravan
184	Denver Public Library, Western History Collection, Otto Perry, #419, RR-419	Rio Grande & Southern locomotive #419 coming up from Rico.
187	Dell A. McCoy Collection, RGS map	RGS map Ridgeway to Durango
188	Denver Public Library, Western History Collection, Robert Richardson, OP-8064	Lizard Head sheep train
194	Courtesy of Navajo Nation Museum, Ralph Leubben	June and Yellowman in Mountain Spring Mine, Rico
226	Rachel Hardwick, Charlie Engle Collection	People's Congressional Church, Rico
242	Courtesy of Navajo Nation Museum	Shiprock
245	Courtesy of Navajo Nation Museum, Milton Snow. NG-52	Navajo mine workers at Kerr-McGee Mine, Cove, AZ
256	Courtesy of Wende Perina Stuart	Rico from Beamis Flat up Silver Creek, Blackhawk Mountain on right
307	PDC Collection	Johnny with his Stevens 12 gauge at a coal tipple near his home, 2009
309	Courtesy of Johnny and Pat	Acid Plant fourteen years after its closure, 1978
319	No photo credit	Author Pat Curran on top of Lizard Head Peak

About the Author

My connection with Rico runs deep. My grandfather, John Patrick Curran and his brothers came west to mine vanadium and carnotite in the early 1900s. They held claims in Naturita, Uravan and in the Carrizo Mountains in Arizona. Thomas F.V. Curran, the wheeler-dealer of the family, worked with Madame Currie to develop a method for processing vanadium into radium, and they say, ran off with what there was of the family fortune.

My Great Grandmother, Mrs. Pauline Sharp, move to Montrose, Colorado, with my grandmother, Stella Sharp. On October 11, 1907 Pauline married Frank P. Batchelder and lived a full life in Noel, Colorado on the Dallas Divide. Grandmother Stella moved on to Telluride and fell in love with John Patrick, who was working his mining claims in Uravan. In 1911, Stella and John Patrick married in Telluride. Stella "True Grit" Curran worked as a cook in the Uravan mining camp and the file shows that she received her full $75 monthly salary for the very month my father, Jack Patrick Curran, was born in 1913.

The Currans moved to Alhambra, California, around 1924. My father, Jack Patrick, married my mother, Rhea Louise Willis, in 1937 and had a long and happy career working for the U.S. Forest Service. We knew it was summer when dad was gone fighting forest fires. My parents retired to Rico in 1971 and dad served a term as Mayor. My mother, Rhea Louise, was a ball of fire. She was active in the Rico Women's Club, and regularly won blue ribbons for her pies at the bake sale. She also had the prettiest garden in town. My brother, Mike, and I have spent part of every year in Rico since the early seventies. Mike is active in the Rico Alpine Club, the Rico Historical Society, and loves to hunt elk and drink at the Hollywood Bar.

As for me, I grew up in the Salinas Valley of California and went to college at U.C. Santa Barbara, followed by four years in the U.S. Navy as an Explosive Ordinance Disposal Diver, including a year in Vietnam. I came home and married Joan Cooper, the prettiest girl in Garden City, New York. I worked as a management consultant in more than forty countries for most of my career. These days I enjoy skiing in the winter and climbing in the summer. Beyond that, I play a shaky game of golf and enjoy my family and my young granddaughters.

Summit of Lizard Head Peak

CPSIA information can be obtained at www.ICGtesting.com
Printed in the USA
267409BV00001B/8/P